THE THAMES TIDEWAY TUNNEL

THE
THAMES
TIDEWAY
TUNNEL

PREVENTING
ANOTHER
GREAT STINK

PHIL STRIDE

The
History
Press

Cover illustrations: *Front:* Lee Tunnel. *Back:* Brickwork arch forming access chamber to the Fleet Main Sewer.

First published 2019
Reprinted 2024

The History Press
The Mill, Brimscombe Port
Stroud, Gloucestershire, GL5 2QG
www.thehistorypress.co.uk

British Library Cataloguing in Publication Data.
A catalogue record for this book is available from the British
Library.

ISBN 978 0 7509 8981 7

Typesetting and origination by The History Press
Printed by TJ Books Limited, Padstow, Cornwall

CONTENTS

A visit with key stakeholders to the Fleet Main Sewer, standing behind an original cast-iron flap valve that was installed as part of Bazalgette's sewer system.

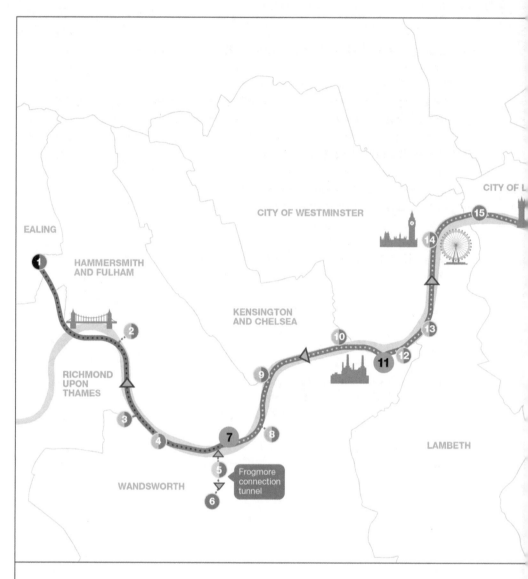

Map key

- ⬤ Main tunnel drive tunnel site
- ⬤ Main tunnel reception site
- ⬤ CSO site
- ⬤ Short connection tunnel drive site
- ⬤ Long connection tunnel drive site
- ◯ System modifications

- ▬ Main tunnel
- ·········· Connection Tunnels
- ▬▬ Lee Tunnel
- ◁ Proposed drive direction
- ······ West works site
- ······ Central works sites
- ······ East works site

- ❶ Acton Storm
- ❷ Hammersm
- ❸ Barn Elms
- ❹ Putney Emb
- ❺ Dormay Str
- ❻ King George
- ❼ Carnwath R
- ❽ Falconbrook

Map labels

EALING

HAMMERSMITH AND FULHAM

CITY OF WESTMINSTER

CITY OF L

KENSINGTON AND CHELSEA

RICHMOND UPON THAMES

WANDSWORTH

LAMBETH

Frogmore connection tunnel

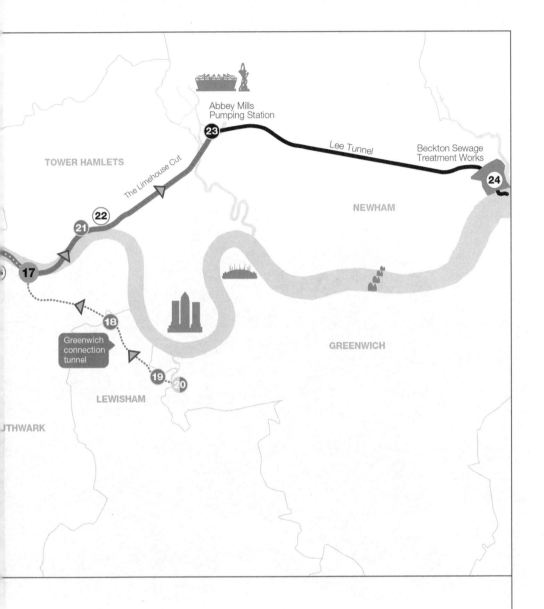

Abbey Mills
Pumping Station

Lee Tunnel

Beckton Sewage
Treatment Works

TOWER HAMLETS

The Limehouse Cut

NEWHAM

Greenwich
connection
tunnel

GREENWICH

LEWISHAM

SOUTHWARK

9	Cremorne Wharf Depot	17 Chambers Wharf
10	Chelsea Embankment Foreshore	18 Earl Pumping Station
11	Kirtling Street	19 Deptford Church Street
12	Heathwall Pumping Station	20 Greenwich Pumping Station
13	Albert Embankment Foreshore	21 King Edward Memorial Park Foreshore
14	Victoria Embankment Foreshore	22 Bekesbourne Street
15	Blackfriars Bridge Foreshore	23 Abbey Mills Pumping Station
16	Shad Thames Pumping Station	24 Beckton Sewage Treatment Works

Wick Lane under construction. (Courtesy Mike Jones)

FOREWORD

BY SIR PETER BAZALGETTE

It's curious that if a major public project goes well, it can be taken for granted. But if there are problems, you hear all about it. The Thames Tideway Tunnel has been very successful thus far, so the usual rule applies. It's all the more welcome, then, to read Phil Stride's painstaking record of this unprecedented piece of civil engineering.

Eminently sensibly, the Thames Tideway Tunnel builds on the legacy and ideas of Sir Joseph Bazalgette's celebrated Victorian drainage system for London (declaration of interest: I'm his great, great grandson). We're still using it five generations on. I don't believe we're undertaking enough schemes today of ambition and vision that will benefit our ancestors. But the Thames Tideway Tunnel is just such a project. Like its Victorian antecedent, it'll allow London to continue to expand as the world's population gravitates towards cities. Phil Stride argues this well here. It's about health *and* wealth.

There's something reassuring – comforting, even – when history repeats itself. The Thames Tideway Tunnel relies on gravity propulsion and it runs from west to east, as did Sir Joe's system. There have been public protests, now as then (Stride sets out a compelling blueprint for a patient and honest, modern consultation). Planning, according to Stride, was the most challenging, as it was for my ancestor when he dug up the whole of London. And the project team has needed to win the support of all major political parties – just as in his time Bazalgette negotiated funds from both a Tory Chancellor (Disraeli) and a Whig one (Gladstone).

For those of you who enjoyed the BBC2 series on the Thames Tideway Tunnel (a hymn to pride, passion and pile-driving), this book now gives you the detailed story of how the 'Super Sewer' was planned, designed, funded and is being executed. A valuable record for all those that follow.

It's not just about public and ecological health, though it is. It goes beyond economic benefit, though that's crucial. This is also about quality of life in our capital, creating more valuable public space. Now that's what I call a legacy project.

Sir Peter Bazalgette
October 2018

LIST OF ABBREVIATIONS

AOD	Above Ordnance Datum (generally, above sea level)
ATD	Above Tunnel Datum (where 0m AOD = 100m ATD)
BLP	Berwin Leighton & Paisner
BTL	Bazalgette Tunnel Ltd
CDM	Construction, Design and Management
CEO	Chief Executive Officer
CFD	Computational Fluid Dynamics
CFO	Chief Finance Officer
CIWEM	Chartered Institution of Water and Environmental Management
CoCP	Code of Construction Practice
CSO	Combined Sewer Overflow
DCO	Development Consent Order
DEFRA	Department for Environment, Food and Rural Affairs
EA	Environment Agency
EaSE	Early Safety Engagement
EPIC	Employee Project Induction Centre
EU	European Union
FFT	Flow to Full Treatment
FROG	Foreshore Recording and Observation Group
GLA	Greater London Authority
GSP	Government Support Package
HMG	Her Majesty's Government
HoTs	Heads of Terms
HSSE	Health, Safety, Security and Environment
HSW	Health, Safety and Wellbeing
IBD	Initial Briefing Document
ICE	Institution of Civil Engineers

IM	Information Memorandum
IP	Infrastructure Provider
ITN	Invitation to Negotiate
ITT	Invitation to Tender
IWRM	Integrated Water Resource Management
M&A	Mergers and Acquisitions
MAB	Metropolitan Asylum Board
MD	Managing Director
MEAT	Most Economically Advantageous Tender
MEICA	Mechanical, Electrical, Instrumentation, Control and Automation
MOLA	Museum of London Archaeology
MoU	Memorandum of Understanding
NAO	National Audit Office
NGO	Non-Governmental Organisation
NPV	Net Present Value
NSIP	Nationally Significant Infrastructure Project
OAWSI	Overarching Archaeological Written Scheme of Investigation
OBC	Outline Business Case
OJEU	Official Journal of the European Union
OSW	Old Sun Wharf
PQQ	Pre-Qualification Questionnaire
R&C	Revise and Confirm
RATS	Residents Against the Thames Sewer
RIDDOR	Reporting of Injuries, Diseases and Dangerous Occurrences Regulations
SAS	Sustainable Active Safety
SCADA	Supervisory Control and Data Acquisition
SBP	Stakeholder Briefing Panel
SIP Regulations	Specified Infrastructure Projects Regulations
SME	Small and Medium Sized Enterprises
SRG	Senior Reference Group
STEM	Science, Technology, Engineering and Mathematics
STW	Sewage Treatment Works
SuDS	Sustainable Drainage Systems
SUDS	Sustainable Urban Drainage Systems
SYR	SaveYourRiverside
TBM	Tunnel Boring Machine
TDS	Tunnel Drive Strategy
THIS	Tideway Heritage Interpretation Strategy

TTSS	Thames Tideway Strategic Study
TTT	Thames Tideway Tunnel
TWUL	Thames Water Utilities Limited
UBS	Union Bank of Switzerland
UCR	Utilities Contracts Regulations
UWWTD	Urban Waste Water Treatment Directive
VDD	Vendor Due Diligence
WACC	Weighted Average Cost of Capital
WWTW	Waste Water Treatment Works (also known as Sewage Treatment Works)
WHO	World Health Organisation
WOOR	Win Only One Rule

'Death's Dispensary', a cartoon that appeared in the Victorian magazine *Fun* in 1866. (Courtesy Mike Jones)

INTRODUCTION

TURNING THE TIDE

THE CAMPAIGN TO BUILD LONDON'S NEW WASTE WATER SYSTEM

Man and boy. That's how long I've worked in the water industry as an engineer and project manager, and nearly all of that time with Thames Water. I remember working, when young, on building sites for my grandfather, who owned a construction business with his brother. They built a lot of the impressive buildings in the Bath area, and I just enjoyed being around it all. In that sense I was always involved with building work growing up, and I loved to see their satisfaction in what they had built.

I got my career break from my mother, though, who worked in education and had an eye for my future. When I was about 15 she introduced me to three family friends in different professions to discuss 'next steps'. One was an accountant, one a solicitor and one a civil engineer. For me there was no choice, the latter meeting propelling me on the career I still love to this day.

The first step was to study civil engineering at Cardiff University. I applied to Thames Water for student sponsorship when I was at school in February 1974 (in fact, initially applying to Swindon Borough Council, given that the Thames Water Authority didn't come into being until 1 April that year). I was one of two such students the company sponsored, then staying with Thames Water until the formation of Bazalgette Tunnel Ltd (known as Tideway) on 24 August 2015. That means I was with the company for forty-one years, in that time having twenty-three jobs (and five in one eighteen-month period). Man and boy!

I am writing this book as a testament to all of those who have worked and strived to make the Thames Tideway Tunnel a success, a project full of superlatives. It is the UK Water Industry's largest ever engineering project, at an estimated cost of £4.2 billion,* which featured the UK's largest ever public consultation and biggest ever planning application. It is also a project where all of those involved are aware of the responsibility being entrusted to them. As Mike Gerrard, former Managing Director of the Thames Tideway Tunnel, says: 'Everyone associated with the project was motivated by the desire to do for our great-grand-children what our great-grandparents had done for us in terms of infrastructure.'

In many ways those years at Thames Water prepared me per-fectly for the role of heading up, and now narrating the story of, the Thames Tideway Tunnel project. The positions I held encompassed every aspect of working for a major water utility, ranging from civil engineering and capital delivery to stakeholder engagement at every level. My roles included being the Head of Operational Control; Head of Capital Delivery; and finally, from April 2008 through to August 2015, Head of Thames Tideway Tunnel for the water utility. Since the creation of Tideway as a separate entity I have been the External Affairs Director, and am now Strategic Projects Director. Throughout this career path I have always had a very close relationship with the Institution of Civil Engineers (ICE), at one time being the UK civil engineering manager of the year (the only UK utility-employed engineer to do so) and ultimately becom-ing a Fellow in 2014. I am also a Fellow of the Royal Institution of Chartered Surveyors. I have also, since 2011, been a member of the Tunnelling and Underground Construction Academy Industry Advisory Panel (IAP), a role that has allowed me to transfer many learnings from Crossrail to the Thames Tideway Tunnel project. I am now chairman of the IAP.

As my years in the water utility sector have progressed, I have always felt a very close association with Sir Joseph Bazalgette, the Victorian engineering giant. I actually saw his work at close quar-ters for the first time when heavily involved in the development of the new station 'F' at the Abbey Mills Pumping Station in Abbey Lane, London. The problem addressed at the time was the amount of sewage going out into Abbey Creek and the River Lee, because there wasn't enough pumping capacity to pump the increasing

---

* In 2014 prices.

flow of waste water to the Beckton waste water treatment works (WWTW). An additional pumping station was therefore needed, to add to what was Bazalgette's largest and arguably greatest piece of above-ground engineering infrastructure, which was a key component of his groundbreaking interceptor sewer system. One of my key roles was therefore to draft the briefing for the consultants bidding to do the design work in the late 1980s. To do that, I had to get down and dirty – quite literally – with the engineering solutions behind his original work. That was when I got my first real appreciation of his genius, and the weight of responsibility that we in our generation have inherited. In fact, when I visited the site in 1987 the original Bazalgette drawings were laid out on dusty tables in a back room.

With all that experience in engineering design, project management and directing, then as head of Capital Delivery at Thames Water, and then being in charge of the Thames Tideway Tunnel project, all threaded through with the pioneering work of Bazalgette, I feel I am uniquely positioned to tell the tale of Tideway to date. Mike Gerrard eloquently describes the nature of this story: 'Trying to make a major project like the Thames Tideway Tunnel happen is the equivalent of waiting until there is a planetary alignment before launching a space rocket.'

Given these challenges, my real aim is to present to the wider world – whether to governments, industry or the wider public – the lessons I have been fortunate to learn in the process of bringing this new project to fruition from early 2008 to today. Additionally, this book acts as a public place of record for all of the activities with regard to the Thames Tideway Tunnel project to date. It also enables me to tell our side of what, while successful, has been an often controversial project. Certainly, being screamed at to one's face at public meetings while trying to explain the rationale behind the programme is character-building!

A final reason for writing this book is with regard to the deserved legacy of all of us who were involved in the Thames Tideway Tunnel project, in the hope that we can encourage others to follow our career paths. As I hope will become clear as you read this book, I love engineering and all things associated with it, with the Thames Tideway Tunnel being the pinnacle of my involvement in many major projects over time. More importantly, however, I also love inspiring others to do so. For me, the most important thing about the privilege of being in a leadership position in my industry, particularly of something as enormous in scope and scale as the Thames Tideway Tunnel, is encouraging people equally to love civil engineering and to achieve their potential therein. This book is their legacy, and also the means by which I hope to inspire a new generation to get involved in our industry, and engineering more broadly.

The key driver behind the first aim of the book, to record our learnings, is set against the widely accepted view that for the vast majority of the inhabitants of the planet the future is urban. In that regard, the World Health Organisation (WHO) says that in 2014, 54 per cent of the world's population then lived in urban areas, with this figure likely to reach at least 66 per cent by 2050. In that timeframe, by 2045, the urban population of the world will pass 6 billion. This will put increasing strain on the ability of governments, both national and municipal, to provide and maintain the infrastructure to support such dense and concentrated settlement, including waste water provision. Therefore, any learnings that can facilitate the engineering solutions that will be required to meet these challenges, especially as they will be largely below ground, and also how to communicate their necessity to the wider public, are vital.

London, of course, presents a superb example of the problems that the growing cities of the future will face. London's population had reached over 8 million by 2012, and yet the waste water network for this mass of humanity is still largely dependent on the Victorian sewerage system of Bazalgette, sometimes updated since but still largely intact and designed to cater for a population of 4 million. This throws into stark clarity why, as I write, Thames Water has to discharge an average 18 million tonnes per year of untreated waste water (combined raw sewage and rainwater) into the tidal River Thames after as little as 2mm of rainfall (these are called combined sewer overflow (CSO) discharges, about fifty such events a year). While the rather unpleasant result of this might not be apparent to the wider public, it certainly is to those who regularly use the river for work and pleasure. As becomes evident as the narrative of the book progresses, a key factor here is that this is well in excess of the 'exceptional circumstances' outlined in the Urban Waste Water Treatment Directive (UWWTD) and is clearly hugely damaging to the ecology of the river. I am often asked what would happen to this volume of discharge if the London Tideway Improvements programme hadn't been initiated, and respond to shocked silence that the figure would rise to an average 70 million tonnes of sewage a year in the river by 2030, with a possible return of the nineteenth-century 'Great Stink of London' as detailed in Chapter 2.

This leads us on to the response to the issue itself, and the focus of the book: the Thames Tideway Tunnel project, the world's first privately financed major sewage overflow tunnel. This is fully outlined in engineering terms in Chapter 6, but I feel it is useful to give a very brief outline here for the general reader to inform the earlier chapters of the book. The programme is the single most important twenty-first-century investment in London's waste water infrastructure, building on the legacy of Bazalgette but with a view to future proofing the city's provision in this regard. The major component is the bored tunnel itself, 25km in length, which, in large part, replicates the course of the River Thames under which it is being constructed (together with the Frogmore and Greenwich long connection tunnels and nine short connection tunnels), taking flows from both sides of the river to a significantly extended Beckton WWTW and with significant new infrastructure at many of the current CSOs through which storm discharges are emitted. To give some idea of scale, for geological and procurement reasons the main tunnel has had to be divided into three sections that are being constructed separately but concurrently:

The West section, from Acton Storm Tanks in the London Borough of Ealing to Carnwath Road Riverside in the London Borough of Hammersmith & Fulham. This section includes the Frogmore connection tunnel.
The Central section, from Carnwath Road Riverside to Chambers Wharf in the London Borough of Southwark.
The East section, from Chambers Wharf to Abbey Mills Pumping Station in the London Borough of Newham. This section includes the Greenwich connection tunnel.

Given this scale, and the focus on the legacy of the Thames Tideway Tunnel experience for all concerned, a Tideway Heritage Interpretation Strategy has been designed to help all stakeholders (whether internal or external) focus on the relationship between the River Thames and the completed tunnels and associated twenty-four sites via legacy commitments to leave a lasting benefit. Indeed, this is one of the key learnings from our project, that one needs to look beyond simply the building of the infrastructure itself with a view to improving the lives of Londoners after construction has been completed.

I would like to thank all those who have made this book possible. Firstly, all of my colleagues at both Thames Water and Tideway who have made this fantastic project realisable: particularly Martin Baggs, former Chief Executive Officer of Thames Water; Mike Gerrard, MD of Thames Tideway Tunnel; and Andy Mitchell, CEO of Tideway, for their unwavering support and encouragement over the years. I would also like to thank Jim Otta of CH2M, whose energy, drive and commitment over many years was critical to our success. Secondly, all those who supported the project, in good times and bad. You know who you are! Thirdly, the opponents, who have served to make the experience more memorable and to whom I hold no grudge. And, of course, my publisher for believing in this work. Others also deserve a specific thank you, for example Richard Aylard, Sustainability and External Affairs Director at Thames Water, and Nick Tennant, former Head of Communications and Public Affairs at Tideway, and my patient proofreader Simon Elliott. Also, a big thank you to Ingrid Lagerberg, Systems Engineering Lead, for diligently ensuring that the text is factually correct. Richard and Nick were hugely supportive throughout the story of our tunnel, and particularly in relation to public meetings and dealing with stakeholders. All have contributed freely and greatly to my wider research, enabling this work on the Thames Tideway Tunnel to reach fruition (and in that regard, Appendix VII lists all of those deserving of a specific thank you for helping take the programme forward at a leadership level).

Finally, of course, I would like to thank my family: my Mum and Dad, Marlene and Ted, for inspiring me to think that anything is possible, my lovely daughters Annie and Janey and – last but not least – my hugely supportive wife Lyn, who enabled my sustained effort over many years to be possible. Thank you all.

Phil Stride
September 2017

1

BACKGROUND

The River Thames

At the heart of the story of the Thames Tideway Tunnel is the River Thames, the watercourse that defines the capital. The river has such an ability to reach and influence people's daily lives that it has become fundamental to the way the whole Thames Tideway Tunnel project has been approached, particularly with regard to its long-term cultural legacy.

The River Thames originally ran through the Vale of St Albans to East Anglia before reaching the North Sea around modern Clacton-on-Sea. This route was scoured clean, however, by the mile-high glaciers of the Anglian glaciation, which by around 450,000 BC had pushed the river broadly to the channel it occupies today.[1] I live close to the 'Goring Gap', which made the change in route possible.

It is the natural force coursing through the heart of the London, a powerful resource that has nurtured all habitation along its length for the entirety of its existence. Even today, long after the decline of the Pool of London and Docklands as two of the principal ports in Britain, it still remains a point of cultural contact between London and the world.

At 346km in length, the River Thames is the longest river solely in England and the second longest in the United Kingdom (the longest being the River Severn at 354km). Rising at Thames Head in Gloucestershire, the river flows west to east through a broad flood plain. Along its length it runs through a variety of other towns and cities before reaching London, these including Oxford (where it is traditionally known as the Isis), Reading, Henley-on-Thames and Windsor. Given this level of settlement, and its location through the heart of England, the River Thames has frequently found itself at the centre of the historical narrative of the islands of Britain. Examples include the Roman conquest in AD 43 when Aulus Plautius led his 40,000 troops across the river to defeat the native Britons near modern Colchester,[2] the campaigns of Alfred the Great against the Vikings,[3] the Wars of the Roses (the murder of the Princes in the Tower took place next to the Thames),[4] the English Civil War with its famous Siege of Oxford,[5] and into the modern era, the river being set alight during the Blitz.[6] Given this heritage, it is no wonder the river was called 'liquid history' by nineteenth-century Parliamentarian John Burns MP.[7]

The tideway of the River Thames, that is the part of the river subject to tides, is the 160km stretch downriver from Teddington Lock at Ham, this featuring three locks and a weir on the southern bank. It includes the Pool of London, Docklands, Thames Gateway and the Thames Estuary. The tidal range of the river in the tideway, that is the difference between high and low tide (caused by gravitational attraction of the moon and, to a lesser extent, the sun) has a substantial rise and fall of 7 metres. It is, of course, this tidal section of the river that gives its name to the Thames Tideway Tunnel and thus the Tideway business. The river has also given its name to three other areas along its length: the Thames Valley, broadly from Oxford to west London; the Thames Gateway, the 70km stretch of the river valley heading downriver from inner east London along both banks; and the Thames Estuary, overlapped by the Thames Gateway, and including the estuary of the River Medway.

Given the amount of urbanisation along its length, and the fact that it runs through some of the driest areas of Britain (even given its broad flood plain), the discharge of the River Thames is comparatively low, with that of the Severn being twice as big, and that with a smaller drainage basin. The low discharge actually exacerbates environmental issues caused by storm sewage being dumped into the tidal river during even the smallest of rainfall events.

The River Thames is the responsibility of the Environment Agency (EA), shared with the Port of London Authority in the tideway section. Expanding its resonance with all of those associated with the river, it has also given its name to many enterprises, industrial and otherwise. Historically, these have included the famous Thames Ironworks and Shipbuilding Company (at Bow Creek, opposite the O2 Arena) which built the world's first all-iron warship, HMS *Warrior*. More recently, the river's name has been found in use among businesses as wide-ranging as Thames Television, Thameslink and, of course, close to my heart, Thames Water and the Thames Tideway Tunnel.

Two significant canals were built in the eighteenth and nineteenth centuries to link the River Thames to other river basins: the Grand Union Canal from London to Birmingham, and the Kennet and Avon Canal from Reading to Bath. Three other canals linking the river to other parts of the country were also initiated: the Wey and Arun to Littlehampton canal (operational through to 1871), the Thames and Severn Canal (in use until 1927), and the Wilts and Berks Canal. Thus we see the River Thames over time reaching out across the country to link far-reaching regions to its own basin, and ultimately the English Channel and the North Sea.

Along its length the River Thames has a number of significant tributaries. These are particularly important to our story as one approaches London, given their historical and current use as an integral part of the waste water network (planned and otherwise). The main tributaries on the approach to London, listed west to east, are as follows:

Running north: River Churn, confluence with the Thames in Wiltshire; River Windrush, Oxfordshire; River Evenlode, Oxfordshire; River Cherwell, Oxfordshire; River Thame, Oxfordshire.
Running south: River Ock, Oxfordshire; River Kennet, Berkshire; Rivers Loddon and Blackwater, Berkshire; Rivers Colne and Misbourne, Surrey; River Wey, Surrey; River Mole, Surrey.

Within Greater London, the tributaries (together with some man-made waterways) are specifically broken down below into those running north and those running south for ease of reference:

Running north: Stamford Brook; Counters Creek; Kilburn; Westbourne; Tyburn Stream; River Fleet; Walbrook; Hounds Ditch; Bow Back Rivers; Blake Ditch; River Lee (and Rivers Rib and Stort); River Roding.
Running south: Beverley Brook; River Wandle; River Graveney; River Effra; River Neckinger; River Peck; Deptford Creek; Ravensbourne; Kidbrooke; River Quaggy.

Once outside of the Greater London area, again moving east, there are three tributaries: River Mardyke, Essex (running north); River Darent, Kent (running south); River Medway, Kent (running south).

Briefing to rowers at the London Rowing Club, Putney, in February 2010.

Further, and clearly of relevance to this book, the famous Victorian interceptor sewers in London created by Joseph Bazalgette are also intimately connected with the River Thames and its tributaries (see Chapter 2). There is a memorial on the embankment, looking up Northumberland Avenue, to Bazalgette and his hard work.

Meanwhile, the River Thames, both in the past and today, is at the centre of the cultural experience of all of those who live in its vicinity. The river is navigable along the vast majority of its length, this being facilitated by forty-five navigation locks together with their weirs. Sailing, kayaking and canoeing clubs proliferate along the river. The Thames is also famous for its annual rowing events, including the Henley Royal Regatta and The Boat Race between the University Boat Clubs of Oxford and Cambridge between Putney and Mortlake. The river has also been used in two summer Olympic games, in 1908 and 1948.

Along its length, the water of the River Thames varies from fresh-water to almost seawater, this supporting a very diverse set of flora and fauna. Indeed, the story of the river's relationship with the plants and animals along its length is a key feature of this story, given the effect that human-related pollution has had (in the modern era, sewage), this being detailed in the following chapters. The river also has a number of adjacent Sites of Special Scientific Interest along its length, the largest being the 5,449ha North Kent Marshes.

London

The Thames has always been a living river, historically the preferred route of transport of the wealthy and a vital means of earning a living for the working man and woman. This is most evident in London, the largest conurbation along its length. Today, the city is a sprawling and massively diverse community of over 8 million.

Evidence of human habitation in the environs of London along the River Thames dates back into the darkest depths of prehistory, with the earliest evidence of pre-human activity after the Anglian glaciation being Homo Erectus flint tools found in the Thames Estuary at Swanscombe (dating to perhaps 400,000 years ago), and more advanced Neanderthal flint tools from Hillingdon (dating to between 100,000 and 50,000 years ago).[8] Woolly rhinoceros and mammoth bones have been found in association with the latter.

Jumping forward in time to the era of modern human beings, we again have flint tool evidence of hunter-gatherers seasonally hunting in the London region around 28,000 BC, though this ceases again with the onset of the last glaciation to strike the north of Europe. This reached its height (the last Glacial Maximum) in Britain around 18,000 BC, after which humans again returned to the area of modern London and it is from this time that we see evidence of seasonal hunter-gatherers once again, with evidence of Neolithic farming along the banks of the Thames from 4,000 BC. From 2,000 BC, this farming became more advanced and we see increasing evidence of these first pre-Londoners taking control of their environment, for example with Late Iron Age burial mounds, bridges, raised walkways, wharfs and watermills. However, this was not a centre of power or population.

The modern city of London owes its existence to the River Thames, it being founded soon after the AD 43 Roman invasion of Britain, which took place under the Emperor Claudius. A Roman mercantile centre named Londinium, it specifically came into being because of the access the river provided to the interior of the growing province from the Thames Estuary, through which a vast amount of materials of all kinds were imported into Britain. The city developed on two low gravel hills located on the north bank of the river, which marked the first spot at which the Thames narrowed enough moving upriver to be bridged. Recent excavations by the Museum of London Archaeology (MOLA) on the new Bloomberg site off Cannon Street have highlighted the commercial origins of Roman London, with the preserved wood from recovered wax tablets dating between AD 43 and 53 referencing money lending and financial transactions.[9] The Romans also built the first river crossing of the River Thames (the aforementioned bridge), close to the site of the existing London Bridge, and the city was later to become the provincial capital, housing the governor's palace, fortress, amphitheatre, a basilica (law courts) and a forum (marketplace), the latter two together being the largest stone-built structure north of the Alps at the time.

In much the same way as all roads led to Rome in Italy, during the Roman occupation here most roads led to or through London, for example Watling Street (from Richborough on the east Kent coast, through London, and then north-westwards to Wroxeter, Shropshire, where it branches south to Caerleon and north to Chester) is still a feature of the city today. Then, as now, the river was also the feature that attracted industrial activity, for example iron and other metal-working enterprises on the south bank near modern Southwark. Intriguingly in the context of this book, Roman London also featured the city's first fresh and waste water systems, with the associated lead piping and wooden conduits being found in a variety of locations across the Roman-built environment (see Chapter 2), particularly in and around public buildings.

From that point London has ebbed and flowed as it has grown into the great metropolis it is today, spreading from its ancient Roman heart (the City of London, which remains its financial centre) first westwards towards the now West End, and then exploding outwards in all directions to become one of world's leading centres of economic enterprise and culture, with an area (for Greater London) of 1,572km². In that regard, Wordsworth's poem 'Composed Upon Westminster Bridge', written in 1802, gives a great insight into the majesty of the River Thames in London in the early nineteenth century.

Today, London has devolved government in the form of the Mayoral Authority and London Assembly, being split into thirty-three local government districts – thirty-two boroughs and the city. These boroughs range from Harrow in the extreme west, Enfield to the north, Havering in the east and Croydon to the south. The Thames Tideway Tunnel affects fourteen of these riparian London local authorities.

Thames Water

The key player in the story of the Thames Tideway Tunnel and the tideway itself is Thames Water Utilities Ltd. The business had its origins in the Thames Water Authority, founded in 1974 as part of the Water Act of the previous year, which nationalised all of the regional water utilities and water catchment bodies. They were reorganised into ten publicly-owned water utilities. Those incorporated into the new body included the Metropolitan Water Board, responsible for the water supply in London; the Swindon Corporation Water Department; and the Thames Conservancy, responsible for the management of the non-tidal River Thames.

These new organisations proved somewhat cumbersome and when, in 1989, the ten water utilities were privatised under the Conservative Government of Margaret Thatcher, the opportunity was taken to separate out once again the responsibilities for the provision of water supply/waste water services and the management of water catchment areas. Thus, with the creation of Thames Water, the National Rivers Authority (which later became part of the EA) took on responsibility for river and channels management, navigation and regulation, while the newly privatised water utility took on all of the other roles previously managed by the Thames Water Authority. Sir Roy Watts, Chairman of Thames Water, played a major role in the privatisation process. This showed foresight, energy and outstanding leadership. He died following a fall in the River Thames in 1993, with his body being discovered 46m from Westminster Bridge.

I joined Thames Water in 1974, and so had a front row seat as a very local-authority-type organisation evolved into a commercial business. The push for ever more efficiencies led to a drastic reduction in the operational workforce. Thankfully, one aspect that has survived to this day is the public service ethos.

Meanwhile, other organisations relevant to the Thames Tideway Tunnel story were also being established as part of the new national water utilities infrastructure, for example water services regulator Ofwat, who were given a remit to protect the customers of the new private water utilities. Later, in 1990, the Drinking Water Inspectorate was created to monitor water quality and safety. Today there are twelve major water utilities in the UK, these being Anglian Water, Dwr Cymru (Welsh Water), Northern Ireland Water, Northumberland Water, Scottish Water, Severn Trent, South West

Water, Southern Water, Thames Water, United Utilities, Wessex Water and Yorkshire Water. This wider industry is now regulated under the Water Industry Act 1991.

Superlatives abound when describing the vast responsibilities of Thames Water; for example, it supplies 2.6 billion litres of tap water a day to 8.8 million customers. Similarly, and topically for this story, it also treats 4.4 gigalitres of waste water for 14 million customers using 350 Waste Water Treatment Works (WWTWs), including the one at Beckton in east London, which is the largest such treatment works in Europe (and which is also home to the UK's first large-size desalination plant). Thames Water also operates and maintains 100 water treatment works, 30 raw water reservoirs, 288 pumping stations and 235 clean water service reservoirs to ensure a first-class service to its customers, all through the hard work of its 4,500 employees.

The UK's largest water and waste water services company, the business has an extensive geographic area of responsibility which includes much of Greater London, the Thames Valley, Surrey, Gloucestershire, Wiltshire and Kent (this footprint is the home of some 27 per cent of the UK's population).

Central to this story is Thames Water's role in the Thames Tideway Tunnel, which was initiated on 27 July 2006 by then Minister of State for Climate Change & Environment, Ian Pearson MP, who wrote to the company and asked it to develop two alternative tunnel options to deal with the discharges from London's combined sewer over-flows. Fully detailed in subsequent chapters, this was based on earlier work by the Thames Tideway Strategic Study group. The company worked with a wide variety of stakeholders and submitted its report with recommendations in December 2006. In 2007, the Government selected a three-part solution, known as the London Tideway Improvements, involving an upgrade of existing tidal waste water treatment works, the building of the Lee Tunnel and the building of the Thames Tideway Tunnel. The Government asked Thames Water to take the whole programme forward from that point. The Lee Tunnel and the Thames Tideway Tunnel are known as the London Tideway Tunnels. I was Head of London Tideway Tunnels and worked on the Lee Tunnel for two years, ensuring that planning approval was obtained and that the main works contract was successfully procured.

The programme management team for the Thames Tideway Tunnel was mobilised in March 2008 and the project progressed from that point, building to the submission of the largest-ever planning application in UK history in February 2014, after the UK's largest-ever consultation exercise, which was initiated over two phases from September 2010. Planning permission was granted via a Development Consent Order (DCO) for the Thames Tideway Tunnel by the Government on 12 September 2014, with the main works and finance contracts awarded in 2015. Tideway came into being on 24 August 2015, being officially named Bazalgette Tunnel Ltd, with physical construction beginning in January 2016.

Waste Water Infrastructure

Building a waste water system is generally straightforward, providing all of the modelling and planning has been carried out such that the water catchment and all environmental variables are well understood (see detail in Chapter 6). The system comprises the following components:

Small sewers connect houses and commercial premises. At all sizes of sewer, an issue here is whether the sewer is or links into a 'combined sewer'. This is a sewer that in addition to the waste water (sewage) also collects surface water run off (storm flow), therefore dramatically increasing its chances of being affected by rainfall events when the surface water can overload the combined sewer, leading to the discharge of storm sewage to a watercourse. They are a legacy issue (certainly in the case of London, with its old waste water infrastructure) given that in modern urban built infrastructure, the waste water and surface water are generally separated and have separate systems.

These smaller domestic and commercial sewers flow into larger sewers, and ultimately into the major trunk sewers that carry the bulk flow of the waste water (and in the case of combined systems, both waste water and rainwater).

Where possible, sewers of all sizes use gravity to enable the flow of the waste water, but ultimately there are often network pumping stations (also called lift stations) to facilitate the progress of their contents. At the end of a sewerage system there is a terminal pumping station, from where the waste water is pumped to a waste water treatment works, such as Abbey Mills Pumping Station. In these stations, of all sizes, waste water is directed through to an underground sump (wetwell), which is equipped to electronically detect the level of waste water therein. Once it rises to a certain level, a pump (usually in a dry well) is started to lift the waste water through a rising main using a pressurised pipe system, up to a manhole from which the waste water can gravitate further along the sewerage system. This whole process is continually repeated, hence reliability being a key factor in the design and operation of a waste water pumping station, until the waste water arrives at its point of destination, usually a WWTW. The pumping stations are specifically designed to cope with peak flows, for example during wet weather, through the use of additional pumps. It is London's struggle to cope with increases in flow, due to the size of the city today and the limited capacity of the sewerage system, that is the key reason for the development of the Thames Tideway Tunnel project: because of the prolific number of discharges of waste water into the Thames.

Finally, the waste water arrives at the WWTW (or sewage treatment works) where it is treated to remove the polluting load in the influent, this being turned into an effluent that can be returned to the water cycle, usually via a watercourse. The key elements of this are the oxidation of carbonaceous biological matter and the biological nitrification process, which converts ammonia in the waste water to nitrate by using aerobic and autotrophic bacteria.* The amount of ammonia in the final effluent being one of three means of measuring its cleanness, the other two being biological oxygen demand** and the amount of suspended solids.*** The overall process is very effective, and in the modern age it is a truism that the effluent emerging from a WWTW is actually often cleaner than the river it goes into.

There are a number of phases within the WWTW through which the waste water is treated. In the first instance, inlet works screen out solid matter, such as tissue, rags and grit, after which the waste water flows to a primary sedimentation tank to settle out the majority of gross solids such as faecal matter. Depending on the design of the WWTW there are then two options for the next

---

* Aerobic and autotrophic bacteria are organisms that, in the presence of oxygen, take organic and inorganic substances into their bodies and transform them into organic nourishment. They are essential to all life because they are the primary producers at the base of all food chains.

** The amount of dissolved oxygen needed (i.e. 'demanded') by aerobic biological organisms to break down organic material.

*** Small particles that remain in suspension in the effluent.

stage of the process, these being trickling filters to remove any further contaminants, or an activated sludge plant. The former is a fixed-bed, biological reactor operating in aerobic conditions with the waste water being continuously trickled over blast furnace slag or other media. In this fixed-bed process, any remaining organic material is aerobically degraded through the use of biofilms covering the filter media. The latter, as used for example at Thames Water's Beckton and Crossness WWTWs, is a biological process used to oxidise carbonaceous biological and nitrogenous matter (mainly ammonium and nitrogen) and remove any unwelcome nutrients such as nitrogen and phosphorous. The carbonaceous biological matter is removed through the use of an aeration tank where air is injected into the mixed liquor. A zone in the same tank, which is kept in an anoxic condition,* is used to break down the oxidised nitrogen.

The final stage for the trickling filter process involves a humus settling tank to remove any excess bacteria. Meanwhile, for the activated sludge plant, the final stage is a settlement tank to remove the activated sludge, the surplus being used to create methane for power. At the Beckton and Crossness WWTWs the sludge is also dried and then incinerated.

A final note of explanation here is with regard to good practice in relation to the design of WWTWs. These treatment facilities are designed to treat a multiple of the normal dry weather flow (the flow from a sewerage system in dry weather). This is so that when rainwater enters the sewerage system there is adequate treatment capacity for the additional flow. It is also common practice at WWTWs to provide storm tanks so waste water can be stored when all treatment capacity has been utilised. If the capacity of the storm tanks is utilised and excess flow continues to occur, then the WWTWs have to discharge the excess into the watercourse. If that is not the case, however, and the capacity of the tanks is not in danger of being breached, then the stored flow is returned for treatment once the rainfall event has passed. For clarity here, what I have outlined above is entirely different to the discharge from CSOs, which occur when the sewerage system does not have the capacity to transport the waste water through the sewerage system to the WWTWs.

---

* i.e., in the absence of oxygen.

THE "SILENT HIGHWAY"-MAN.
"Your MONEY or your LIFE!"

'The Silent Highwayman', a cartoon published in *Punch* magazine in 1858.
(Courtesy Mike Jones)

2

THE GREAT STINK: LONDON BEFORE AND AFTER BAZALGETTE

Early Drinking and Waste Water Provision in London

The provision of waste water services in London has always run hand in hand with the provision of fresh water, with the supply of the latter dating back to the Roman period when clay pipes carried water from the Walbrook to the city's baths (for example those on Upper Thames Street, Huggin Hill and Cheapside) and public conveniences. By the medieval period these had long fallen out of use and drinking water was being drawn from the Thames and its tributaries (setting the scene for the future problems), together with natural wells.

For most of the citizens of London, the process of gathering water from these sources involved carrying their own supply, though the wealthier members of society employed waters carriers (who in 1496 formed themselves into a guild called The Brotherhood of St Cristofer of the Waterbearers). However, growing demand caused some of the smaller tributaries to dry up, and by the early thirteenth century water piping began to appear once again, this time in the form of hollowed-out elm tree trunks, sandstone, lead and (again) clay. Fresh water provision of this nature increased in scale as the century progressed, and in 1236 Gilbert de Sandford was granted 'liberty to convey water from the town of Tyburn by lead of pipes into the city'. The water so supplied was carried to the city by a great conduit of such lead pipes along the line of

Oxford Street and Holborn through to Cheapside, with modern Conduit Street marking its route today.[1] Other such conduits were erected as the city grew over the next few centuries, for example that now marked by the route of Lamb Conduit Street after a benefaction by Sir William Lambe in 1577.

A further major development to satisfy the ever-growing demand for fresh water was the building of the New River from 1613 by Sir Hugh Myddleton, Member of Parliament for Denbigh. This was a 61km waterway designed to bring water from a spring located at Amwell in Hertfordshire to Sadler's Wells north of Clerkenwell. In order to give the project a firmer legal and financial structure, the New River Company was set up and incorporated in 1619 by Royal Charter. It was one of the first companies in the world. The Oak Room at New River Head, Rosebery Avenue, houses the original boardroom – it is arguably the oldest remnant of the UK water industry, and a room where I had my Thames Water leaving meal. A century later, in 1723, the Chelsea Water company was established to take water from the River Thames's north bank, this to be followed by the creation of six other such water provision companies over the next 122 years. These were the West Middlesex, Grand Junction Waterworks, East London, South London, Lambeth and Southwark companies.

As the fresh water provisions grew, so did the waste water systems – although they did lag somewhat behind. Roman London had an ad hoc sewerage system that provided a service to specific, often elite or public, buildings and latrines. An example is provided by two 1998 excavations on the terraced ground north of the River Thames to the south of Cornhill. Here, a third-century stone building has been excavated that includes a subterranean drainage culvert designed to carry waste water south to the river. Meanwhile, at 13–21 Eastcheap, some stone-built structures dating to after a major fire in London during the reign of Hadrian in the early second century have been found with timber drains. Yet another specific example can be found at the Roman amphitheatre, wonderfully exposed in part beneath the current Guildhall Gallery, where the wooden drains under the actual arena have been exposed (though in this case, possibly for grimmer uses than the removal of waste water). More broadly during the Roman occupation, though, and certainly into the Medieval period, most of London's sewers were open ditches running into the River Thames and its tributaries.

The first visibility we have of post-Roman waste water provision in London and elsewhere is an Act of Parliament designed to deal with the matter in 1427, after 'a year of unseasonal weathering'. This called for a number of commissioners of sewers to be recruited over the ensuing ten years (on a national scale). This Act was renewed five times over the next century, but the provision afforded by these Acts always fell short of the requirement. To that end, Henry VIII passed the 'Bill of Sewers' in 1531, due to the 'gret wyndes and fluddes' of the preceding year.[2] In addition to facilitating the appointment of commissioners with greater powers to run seven commissions, this also decreed that homeowners had responsibility for keeping the section of sewer in front of their properties clean. Again, though, things fell short of requirements with regard to London, with open ditches and the River Thames and its tributaries continuing to be used for waste water and waste disposal. A 1605 Act was eventually passed giving the commissioners in the city powers over all of the:

> walls, ditches, banks, gutters, sewers, gotes [conduits], causeys, bridges, streams and watercourses, navigable or not, within the limits of two miles, of and from the City of London, which waters have their course and fall into the River Thames.[3]

The reference here to the powers being granted with regard to unnavigable watercourses is significant, as it is indicative of a city growing rapidly. Again this Bill ultimately fell short, however, this now being a regularised pattern, with more powers added to the commissioners' brief by an Act in 1690, which gave them responsibility for any drains and sewers built since the 1660 Act within the cities of London and Westminster.

Even if these developments improved matters, however, by 1827 things were in a poor state again. A pamphlet was written by the journalist John Wright called 'The Dolphin, or Grand Junction Nuisance, Proving that Several Thousand Families in Westminster and its Suburbs are Supplied with Water in a State Offensive to the Sight, Disgusting to the Imagination and Destructive to Health'. This highlighted pollution in the River Thames and its tributaries caused by sewage and industrial effluent, specifically targeting the Grand Junction Waterworks, which supplied drinking water to Wright's Regent Street home. In the pamphlet, distributed to Parliament, he described the water used at his breakfast table as:

a fluid saturated with the impurities of 50,000 homes – a dilute solution of animal and vegetable substances in a state of putrification ... offensive to the sight ... [and imagination and health]

Wright, an associate of social reformer William Cobbett, dedicated the pamphlet to radical firebrand MP Sir Francis Burdett, who championed the issue, and this led to a Royal Commission on the matter in 1828. This found that: 'many of the complaints respecting the quality of the water are well founded; and that it ought to be derived from other sources than those now resorted to.'[4]

The commission's report led to a Parliamentary Select Committee inquiry, on whose recommendation leading national engineer Sir Thomas Telford, by then 71, was appointed to look into new means of providing the capital with fresh water. His report proposed to avoid the River Thames as a source of new water supply and instead use three new and unpolluted sources for water provision that would be uncontaminated by the city's waste water: the River Ver at Aldenham, the River Wandle at Beddington and the upper reaches of the River Lee. This proposal had an estimated cost of £1,177,840 16s 5d. A further Select Committee of 1834 considered the matter but it seems no action was taken.

A new attempt to legislate against the increasingly polluted river was made in 1839, when a Bill was introduced in Parliament by Benjamin Hawes MP and Henry Ward MP aiming to establish a Metropolitan Court of Sewers, with wide-ranging powers to improve the quality of the water supply. The Bill failed, however, the reasons being lost to us. Later legislation did take place in the form of the 1844 Metropolitan Buildings Act and the 1846 Nuisances Removal and Diseases Prevention Act, but the main issue preventing their improving water quality seems to have been a lack of enforcement.

As can be seen, the general trend of the debate about the provision of both drinking and waste water services in London was now becoming part of the wider debate on public health, and the next actor to grace the stage was Edwin Chadwick, another great nineteenth-century social reformer and leading member of the Sanitary Movement. Concerned about the living conditions of the poor in London and elsewhere, he had made his reputation through his contributions to the groundbreaking 1834 Poor Law Amendment Act. With regard to his concerns of public health, Chadwick had famously proclaimed in front of Parliament in 1846 that 'All smell is disease', this reflecting his (then common) view that disease (for example the frequent and fatal urban out-breaks of cholera)* was caused by polluted air, referred to then as 'miasma'. In 1849 he became the commissioner of the new General Board of Health, with a key priority being the modernisation of London's ad hoc, rapidly ageing sewerage system. Progress was actually made in this regard in the late 1840s, as he was able to force through the banning of private cess pits in the capital, with all waste water now having to be connected to the sewerage system. More importantly, following his appointment a number of the older brick sewers were replaced with new, round earthenware pipes (based on an ancient Roman design and more expensive than the designs they replaced) set at a gradient to use gravity to assist their flow. While this did amount to progress, reducing the above-ground 'miasma' given the natural flow away from urban areas, they did still empty into the River Thames. This exacerbated the existing issue of the polluted river, which continued to pro-duce a prodigiously bad smell, particularly in the summer. This had already been widely recognised, for example with *Punch* in 1848 publishing its famous poem on the matter:

Filthy river, filthy river,
Foul from London to the Nore,
What art thou but one vast gutter,
One tremendous common shore?

All beside thy sludgy waters,
All beside thy reeking ooze,
Christian folks inhale mephitis,
Which thy bubbly bosom brews.

All her foul abominations
Into thee the City throws;
These pollutions, ever churning,
To and fro thy current flows.

---

*   The outbreaks of cholera were actually caused by enterotoxin-producing strains of the bacterium Vibrio cholera, which had entered the drinking water supply through waste water pollution. Even though Dr John Snow had correctly linked contaminated water supply as the means by which the disease was communicated in 1849 (and proved his case during the 1854 Broad Street outbreak), the link to it being spread by the 'miasma' remained until the epidemic of 1866.

And from thee is brewed our porter –
Thee, thou gully, puddle, sink!
Thou, vile cesspool, art the liquor
Whence is made the beer we drink!

Thou too hast a conservator,
He who fills the civic chair;
Well does he conserve thee, truly,
Does he not, my good Lord Mayor?

Ultimately, Chadwick was unsuccessful in his wider ambitions because, though his campaigning brought the polluted nature of the river to the attention of a wider public audience, he was becoming unpopular politically. Additionally, his vociferous agitation towards the water companies meant that they would never offer their support, vital if an ultimate solution was to be found.

The next serious attempt to clean the river and its tributaries was made through the Metropolitan Sewers Act of 1848 (following an 1847 Royal Commission chaired by Chadwick and Lord Morpeth) that established one Metropolitan Commission of Sewers, amalgamating those dating back to Henry VIII's time. Each commission lasted two years, with six being held from 1848 to 1855, though again little progress seems to have been made. Charles Dickens himself intervened in the debate in April 1850, noting in his campaigning journal *Household Words* the complacency of the representatives of the water companies.[5] The only exception here appears to have been the Chelsea Water Company, which back in 1829 had introduced a sand filtration system for its drinking water before it was piped to people's houses. It would be another sixty years, however, before the work of Louis Pasteur showed how this system actually worked in practice, the sand proving to not only be a physical but also – and crucially – a biological barrier to the impurities in the water before it was so filtered.

Next, the Metropolis Water Act 1852 was passed by Parliament, this making it illegal from 31 August 1855 for any of the water companies to extract domestic-use water from the tidal River Thames. The companies also had to filter any water extracted for domestic use from 31 December that year. Things still seem not to have improved, however.

The next great character of the age to consider the issue of impending environmental disaster was Michael Faraday, the scientist famous for his contributions to our understanding of electromagnetism, electrochemistry and sanitation, and one of the leading intellectual minds of the time. Disgusted by the polluted waterway, Faraday conducted a series of experiments that would allow him to test the overall quality of the water. He collected his findings and decided to voice his worries publicly. On 7 June 1855, he wrote the following letter to Parliament and *The Times* newspaper. At the time, *The Times* was the most influential newspaper in the country:

Sir, I traversed this day by steam-boat the space between London and Hungerford Bridges between half-past one and two o'clock; it was low water, and I think the tide must have been near the turn. The appearance and the smell of the water forced themselves at once on my attention. The whole of the river was an opaque pale brown fluid. In order to test the degree of opacity, I tore up some white cards into pieces, moistened them so as to make them sink easily below the surface, and then dropped some of these pieces into the water at every pier the boat came to; before they had sunk an inch below the surface they were indistinguishable, though the sun shone brightly at the time; and when the pieces fell edgeways the lower part was hidden from sight before the upper part was under water. This happened at St. Paul's Wharf, Blackfriars Bridge, Temple Wharf, Southwark Bridge, and Hungerford; and I have no doubt would have occurred further up and down the river. Near the bridges the feculence rolled up in clouds so dense that they were visible at the surface, even in water of this kind. The smell was very bad, and common to the whole of the water; it was the same as that which now comes up from the gully-holes in the streets; the whole river was for the time a real sewer. Having just returned from out of the country air, I was, perhaps, more affected by it than others; but I do not think I could have gone on to Lambeth or Chelsea, and I was glad to enter the streets for an atmosphere which, except near the sink-holes, I found much sweeter than that on the river. I have thought it a duty to record these facts, that they may be brought to the attention of those who exercise power or have responsibility in relation to the condition of our river; there is nothing figurative in the words I have employed, or any approach to exaggeration; they are the simple truth. If there be sufficient authority to remove a putrescent pond from the neighbourhood of a few simple dwellings, surely the river which flows for so many miles through London ought not to be allowed to become a fermenting sewer. The condition in which I saw the Thames may perhaps be considered as exceptional, but it ought to be an impossible state, instead of which I fear it is rapidly becoming the general condition. If we neglect this subject, we cannot expect to do so with impunity; nor ought we to be surprised if, ere many years are over, a hot season give us sad proof of the folly of our carelessness.[6]

Faraday's warning that the Thames needed to be purified was generally ignored by Parliament. Only three years later, his prediction of the effects of a hot season came true.

The Great Stink of London

Faraday's 'hot season' did indeed occur, three years later in 1858, the hottest summer then on record. In what became known as the 'The Great Stink of London', the stench from the river in the heat effectively brought all activity along the banks of the Thames to a halt. Most importantly in terms of the matter being tackled, it affected the running of the country in that the overpowering smell and atmosphere prevented the operation of the Houses of Parliament. Disraeli himself, the future Prime Minister and then Leader of the House of Commons, is recorded exiting the Chamber of the House of Commons with a handkerchief over his nose complaining of the smell from the 'stygian pool' that the River Thames had become. With regard to the disgusting nuisance, *The Times* wrote in an editorial on 18 June 1858:

> What a pity it is that the thermometer fell 10 degrees yesterday. Parliament was all but compelled to legislate upon the great London nuisance by the force of the stench. The intense heat had driven our legislators from those portions of their buildings which overlooked the river. A few members, indeed, bent upon investigating the matter to its very depth, ventured into the library, but they were instantaneously driven to retreat, each man with a handkerchief to his nose. We are heartily glad of it.[7]

By this time the Metropolitan Board of Works had been established through the Metropolitan Management Act of 1855 to supervise all of the public works across the capital. However, when, in the summer of 1858, MPs debated the problem on the same day as the above article appeared in *The Times*, many were highly critical of the board's

Michael Faraday presenting his card to Father Thames. (*Punch*, July 1855)

attempts to rectify the problem. This debate was widely reported in the media, for example in *Punch* again, the *Illustrated London News* and *The Observer*, with the *Journal of Public Health and Sanitary Review* reporting grown men being struck down by the smell. Ultimately the *City Press* reported that: 'Gentility of speech is at an end – it stinks: and whoso once inhales the stink can never forget it and can count himself lucky if he lives to remember it.'[8]

43

The events of early June brought matters to a head with a general feeling across Parliament that 'something must be done', overcoming the financial concerns that had killed off previous attempts to tackle the problem. On 15 July, Disraeli introduced the Metropolis Local Management Amendment Act, which was designed to extend the powers of the Metropolitan Board of Works to enable them to properly tackle the purification of the River Thames. This became law eighteen days later, with the board empowered to borrow £3 million (the borrowing of a further £1.2 million was approved in 1863), all guaranteed by the Treasury, and with the power of Parliament to veto any plans being removed. The money was to be paid for by a 1½ 'old pence' levy on the business rates. It is at this point that Joseph Bazalgette enters centre stage to save the capital from environmental and financial disaster.

Sir Joseph Bazalgette

Born in 1819 at Hill Lodge, Clay Hill near Enfield in London, Joseph Bazalgette was the son of retired Royal Navy captain Joseph William Bazalgette. His initial engineering experience was on the railways, and he was then articled to the leading early-to-mid-nineteenth-century engineer Sir John MacNeill, with whom he gained much experience, often abroad, in skills such as land reclamation and drainage. This allowed him to set up his own engineering consulting practice in London in 1842. Marrying in 1845, Bazalgette's work at the time was still heavily focused on the rapidly expanding railway network, and he suffered a nervous breakdown in 1847 due to his intense workload.

On his recovery, he was appointed assistant surveyor to the Metropolitan Commission of Sewers in 1849, becoming its engineer in 1852 following the death of his predecessor (from 'harassing fatigues and anxieties'). As waves of cholera struck the capital in the late 1840s and early 1850s, and with the River Thames increasingly toxic to Londoners, Bazalgette generated a number of potential solutions to the problem but all suffered for a lack of funding. Many thought his proposals were far too expensive, interestingly a common theme with the Thames Tideway Tunnel. His efforts were recognised by his engineering peers, however, even if not by politicians, and in 1856 Isambard Kingdom Brunel championed him to become chief engineer of the new Metropolitan Board of Works, which had replaced the commission. He started the job on 25 January that year on an annual salary of £1,000, and was thus in position when the crisis peaked in 1858, enabling him to take advantage of the newly found political backing to solve the problem. His engineering marvel is detailed in full below, marking him out as the leading specialist large-scale project engineer of his generation. It is no surprise that, from that point on, he was the first port of call for any new waste water drainage scheme across the country, with those he worked on in some capacity including:*

Epsom (1858)	Northampton (1871)
Luton (1858)	Skipton (1871)
Netley (1859)	Birmingham (1872)
Feltham (1862)	Scarborough (1872)
Shrewsbury (1862)	Dorking (1872)
Bristol (1863)	Croydon (1873)
Cheltenham (1863)	Beckenham (1873)
Weston (1865)	Norwich (1873)
Oxford (1865)	Farnham (1873)
Hastings (1866)	Norwich (1874)
Folkestone (1866)	Hampstead (1874)
Oxford (1866)	Margate (1874)
Cambridge (1866)	Ruabon (1875)
Glasgow (1868)	Torquay (1875)
Herne Bay (1869)	Portsmouth (1882)

For some of these consultancies he adjudicated between differing competing solutions to a requirement; for example in Portsmouth, where thirteen schemes were submitted to meet the requirement. He also provided consultancy abroad on waste water management, for example at Port Louis in Mauritius (he sent his son

---

* *After Sir Joseph Bazalgette: Civil Engineering in the Victorian City.* Institution of Civil Engineers, 1991.

Edward there to carry out the surveying) and in several European cities, including Budapest.

Within his remit at the Metropolitan Board of Works he was also heavily involved in the three new embankments of the River Thames (fully detailed below in the context of his waste water scheme) and also in the review of the bridge crossings of the River Thames. The context with regard to the latter was a Select Committee Report of 1854 that had lamented the traffic problems involved in crossing the River Thames, an issue the board inherited. The board later moved to buy the principal river crossings from the private companies that owned them, and then free them from any tolls (mainly through the Metropolis Toll Bridges Act 1877). The twelve bridges in question included those at Hammersmith, Putney, Wandsworth, Battersea, Albert, Chelsea, Vauxhall, Lambeth and Waterloo. It was to Bazalgette that the board turned in order to review their condition prior to purchase, which allowed them ultimately to acquire the twelve for £1.5 million (this beginning in 1869). His work didn't end there in this regard, however, and after much repair work across the twelve the decision was taken to replace those at Putney, Hammersmith and Battersea with new designs by Bazalgette. These bridges remain London icons to this day. He was also involved in the design of the infrastructure and systems for the Woolwich Ferry.

In later life Bazalgette also provided further consultancy services on matters related to the railways. He also worked with the Rochester Bridge Trust regarding the Rochester and Maidstone crossings of the River Medway, based on his work with the board in London, recommending that a new bridge be built at the latter location. He designed this structure, which was given the go-ahead in 1877, the first pile being driven into the riverbed in October that year. It opened with great fanfare in 1879.

Throughout this period he remained the chief engineer of the new Metropolitan Board of Works, overseeing the creation of his famous interceptor sewer system. He died in 1891 aged 72 years, with the Institution of Civil Engineers reflecting in their condolences that:

> his life had been given to considerations affecting public health and welfare in all the large cities of the world, and his works, as the engineer for many years of the Metropolitan Board of Works, will ever remain as monuments of his skill and professional ability.

He was president of the Institution of Civil Engineers from December 1883 to December 1884.

Bazalgette's Remarkable Engineering Solution

Bazalgette began his research into how to solve the problem of London's polluted rivers and drinking water supply as soon as he took up his post in January 1856, bringing with him three assistants from the 1855 commission to help. He set himself six questions that needed to be addressed if he was to succeed:

- At what point in the tide of the River Thames should waste water be discharged?
- With regard to the interceptor sewers he was planning to use, what would their minimum fall be?
- In a twenty-four-hour period, what was the quantity of waste water to be managed and its flow pattern?
- Should the sewers be designed to carry rainfall too (the combined sewers detailed in Chapter 1)?
- By what criteria should the size of the sewers be determined?
- What were the best types of pumps and pumping engines for lifting waste water?

Based on the answers, Bazalgette designed his scheme, which he presented to Parliament in the spring of 1858, though as detailed above it took the 'Great Stink of London' to actually provide the momentum to enable the required legislation to pass through the House of Commons. Bazalgette had concluded that it would be too complex and expensive to construct separate waste water and rainwater systems. A combined sewer system was thus chosen. Work began on what became known as the Main Drainage (the interceptor sewers, pumping stations and outfall works) on 31 January 1859, with construction being carried out by a number of contractors who competed for the work through competitive tender. From that point the operation grew in scale, such that by 1865 his department at the Metropolitan Board of Works comprised the three original assistants, twenty draftsmen, five engineer's clerks, an engineer's accountant, two surveyors and up to fifty-nine clerks. The work was largely completed in that year, with the main elements of the drainage system being opened at an extravagant ceremony at Crossness Outfall Works by the Prince of Wales on 4 April, where he toasted: 'the eminent and skilful engineer, Mr Bazalgette'.

Despite this opening, construction was still being completed up to the early 1870s, particularly to the north of the River Thames where delays were the result of the ambitious plans to incorporate the Northern Low Level Sewer into the new Victoria Embankment. This was a part of a long-standing desire by the then Metropolitan Board of Works to embank and landscape the Victoria, Albert and Chelsea embankments. Each embankment required a separate Act of Parliament. Using the momentum and funds freed up by the new Main Drainage scheme, this project was realised under Bazalgette, with the Victoria Embankment incorporating not only the Northern Low Level Sewer but also public gardens, roads and the underground railway. This line, adjacent to Blackfriars Tube station, goes through the top section of the Fleet Sewer. I have often placed my hand on the support girder and felt the vibration of the train. This was opened in July 1870, again by the Prince of Wales, and ran for 2km from Westminster to Blackfriars Bridge. Meanwhile, the Albert Embankment on the south bank of the river was opened in May 1868, running from Westminster Bridge to Vauxhall Bridge. Finally, the Chelsea Embankment was completed during 1874. Overall, the three embankments reclaimed 52ha of new land. At the time, Bazalgette received many plaudits for this work, though he himself said that the Main Drainage scheme was actually more difficult.

Bazalgette's design tackled the problem of the polluted river at source, diverting the waste water far downriver to the Thames Estuary and so avoiding the danger of polluting the river as it flowed through the capital. The outfall sites at Beckton and Crossness were chosen so that, if the balancing tanks were discharged on an incoming tide (they were normally discharged on an outgoing one), the contents would never reach Westminster. He used an extensive underground waste water system, his innovative concept featuring five main interceptor sewers running west to east (in total nearly 160km in length), some of which 'intercepted' the 'lost rivers' of London such as the Fleet and the existing waste water system (such as it was). Of the five interceptor sewers, three were to the north of the river, these being:

A channel at the end of the Fleet Main Sewer, and the flap valve allowing discharge to the River Thames through the northern abutment of Blackfriars Bridge.

The Northern High Level Sewer, starting at Hampstead Hill, then running through Kentish Town, Stoke Newington and under Victoria Park before joining the Northern Outfall Sewer at Wick Lane, Hackney. The flow had enough head to gravitate to Beckton Outfall Works.

The Northern Middle Level Sewer, starting near Notting Hill before joining the Northern Outfall Sewer at Wick Lane, Hackney. The flow had enough head to gravitate to Beckton Outfall Works.

The Northern Low Level Sewer, starting to the west of Counters Creek in Chelsea before being pumped at Abbey Mills Pumping Station into the Northern Outfall Sewer downstream of Wick Lane, Hackney. Part of this sewer was incorporated into the Chelsea and Victoria Embankments as detailed above.

The Northern Outfall Sewer, a major trunk sewer, then carries the flows from the three interceptor sewers to Beckton Outfall Works. An additional mid-level and an additional low-level sewer were added to this system around thirty years after Bazalgette's network was completed.

Meanwhile, to the south of the river were the two further interceptor sewers, these being:

The Southern High Level Sewer, which begins at Herne Hill, then travels under Peckham and New Cross before joining the Southern Outfall Sewer at Deptford. The flow had enough head to gravitate to Crossness Outfall Works.

The Southern Low Level Sewer, starting at Putney and running through Battersea and Vauxhall and then under the Old Kent Road and Bermondsey before joining the Southern Outfall Sewer at Deptford. The flow was pumped to Crossness Outfall Works via the Greenwich Pumping Station (the Abbey Mills Pumping Station equivalent on the south side of the river), which lifts the flow from this sewer, it being immediately pumped again by the Inlet Pumping Station on arrival at Crossness Outfall Works.

These sewers both flowed into the Southern Outfall Sewer, a major trunk sewer carrying the two combined waste water flows from Deptford, then under Greenwich, Woolwich and Plumstead, and finally through the Erith Marshes to the Crossness Outfall Works in south-east London. Both the Beckton and Crossness Outfall Works consisted of holding tanks which filled with waste water. Becton and Crossness were known as 'Outfall Works' as they only provided holding tanks to allow contents to be discharged on the outgoing tide. They would later be known as WWTWs (or sewage treatment works) when treatment facilities were added.

The new interceptor sewers were fed by 720km of existing and new main sewers, which in turn conveyed the waste water from 21,000km of smaller localised sewers. The design capacity of the new system in terms of population was 4 million inhabitants in the capital, at the time thought to be a realistic top end figure, though of course this has proved not to be the case. The scale of Bazalgette's interceptor sewer system is self-evident in the statistics of its construction. For example, it required 318 million bricks, 670,000 cubic metres of concrete and 2.7 million cubic metres of excavated earth to build. A particular innovation was the use of Portland cement to strengthen the sewers, which are still in excellent condition today. Bazalgette's interceptor sewers in London were the largest civil engineering project in Britain in the nineteenth century. The scheme as initially proposed did not extend as far west, on the north of the river, as Hammersmith and Fulham. This area at the time was a market garden for the capital. A number of doctors and local dignitaries petitioned Parliament, however, highlighting the health risks of the area not being connected to the interceptor sewers. They were concerned that the area would become a polluted marsh. So having scheme protestors in the London Borough of Hammersmith and Fulham is not a new phenomenon!

The protestors won the day and the Northern Low Level Sewer was extended. As there was not an adequate gradient to allow the extension of the gravity system, a 'new' pumping station was built at Pimlico called the Western Pumping Station, with its very famous and recognisable chimney. In the design, gravity was used to allow the waste water to flow eastwards, though at Pimlico, Abbey Mills, Greenwich and Crossness the pumping stations detailed were built in order to lift the waste water and so ensure sufficient flow for the system to work. Abbey Mills Pumping Station, often known as the 'Cathedral of sewage', is one of the most important industrial heritage sites in the country. Improvements to Bazalgette's system since that time, in addition to the two extra interceptor sewers north of the river, have largely concentrated on waste water treatment provision at Beckton and Crossness with the aim of substantially reducing pollution in the Thames Estuary and North Sea.

Northern Outfall Sewer, 1902.

Bazalgette interceptor sewer junction at Charlton, showing fantastic workmanship and condition of brickwork.

Eltham Relief Sewer, which was constructed in 1939 to supplement Bazalgette's interceptor sewers.

3

THE NEW WASTE WATER CRISIS

Despite major attempts to upgrade the Bazalgette interceptor sewer system in the late nineteenth and twentieth centuries, it still provides the backbone of the waste water network for London. This is the real starting point of the Thames Tideway Tunnel story, with the realisation that as the twenty-first century progressed the growth of the capital, with its modern day population of 8 million as of 2012, was far outstripping the ability of a sewerage system designed to cope with a population of 4 million. This situation has been compounded by the fact that the Bazalgette system features combined sewers (as detailed in Chapter 1), meaning that its inability to cope with large volumes of combined waste water and surface water run off. During rainfall events, this most often manifests itself in untreated sewage being deposited directly into the tidal River Thames and its tributaries through the fifty-seven CSOs, the vast majority of which were constructed as part of Bazalgette's Main Drainage system.

The Global Trend for Urban Living

The United Nations (UN) Department of Economic and Social Affairs (DESA) said, in a 2014 revision report on the world's 'urbanisation prospects', that at that time 54 per cent of the world's population lived in urban areas (totalling 3.9 billion, up from 746 million in 1950). Such urban areas are defined as regions of human settlement featuring high population densities and related supporting infrastructure, usually being associated with cities, towns, conurbations and suburbs. This UN report said that its

A Thames21 volunteer clearing sewage-derived litter, including plastics, from the Putney combined sewer overflow in the southern abutment of Bazalgette's Putney Bridge.

54 per cent figure was expected to rise to 66 per cent by 2050, with projections showing this would add an additional 2.5 billion people to the world's urban populations by that time, again all requiring supporting infrastructure.

The UN report stated that 'mega-cities' (cities with a population of over 10 million) were increasing in number. In that regard, in 1990 there were ten such mega-cities, which housed 153 million people, just under 7 per cent of the global urban population. By 2014 there were twenty-six, housing 453 million people, or around 12 per cent of the world's urban population. Of these mega-cities, sixteen are in Asia, four in Latin America and three each in Europe and Africa. By 2030 the UN predicted there would be forty-one mega-cities. These facts are very instructive, given the above prediction about the growth in urban living through to 2050 in the UN report and the required infrastructure to support this development.

The UN report additionally said that smaller cities also featured prominently in the make-up of the world's urban centres, and also in the expected growth of the urban population through to 2050. Today, almost half of the 3.9 billion urban dwellers live in comparatively small settlements that have fewer than 500,000 inhabitants, many of them the fastest-growing cities in the world.

The implications of all of this change, with the dramatic increase from rural to urban living, are profound. With the world's total urban population expected to exceed 6 billion by 2045, the fact that much of this expected growth will take place in developing countries means these countries will face many challenges in providing for the needs of their growing urban populations. This will include housing, transport, energy, education, employment, healthcare, and of course engineering infrastructure, including adequate and sustainable clean water and waste water provision.

The UN report noted with regard to all of the above trends that successful urban planning will be the prime requirement to ensure the success of urban settlements of all sizes as they grow so quickly. It adds that, providing such growth is well managed in a sustainable way, cities will continue to be beneficial in that they offer opportunities for economic development. Topically, the report also says that providing water and sanitation for an urban population is usually cheaper and less environmentally damaging than providing similar levels of services to a typically dispersed rural population.

It is in the context of all of the findings of the above UN report on 'urbanisation prospects' that we at Tideway are keen to share our experiences, aiming to help inform the sustainable growth of the world's urban environment.

London's New Waste Water Crisis

London represents one of the best examples of how the mistakes of the past through a lack of well-managed, sustainable planning (even with the best intentions) have presented a city with a major legacy issue – in this case, waste water network provision. No matter how well the Bazalgette interceptor sewer system was designed and built, the politicians and engineers of the day could have no idea of how rapidly the city would grow in just over a century, leaving it with a system designed to cope with the waste water of 4 million people having to cater for over 8 million (and with the Office for National Statistics predicting that this figure will increase to 10.5 million over the next twenty years). The issue facing London's sewerage system as the century progressed was therefore one of capacity, and nothing to do with the condition of Bazalgette's system, which remains in excellent condition to this day.

This capacity issue is exacerbated by the fact that Bazalgette's sewers are a combined system, carrying both the waste water and rainwater in the same pipes. Thus, in rainfall events of any note, the volume of waste water overwhelms the normal operation of the system, which then discharges the overflow through one of the CSOs into the tidal River Thames.

A further issue is the gradual concreting-over of the open spaces of the capital in recent decades. This has occurred for a variety of reasons, for example the dramatic rise in the owner-ship of cars. Given the lack of space in London and its suburbs for garages to house the cars (as many of the former were built before the rise of the latter), many front gardens in the city have been concreted over to provide off-road parking. One only has to drive down a suburban street in the early evening in London to see the large number of houses with multiple cars nestling in front, on such concreted-over drives. This means that rainwater that would previously have soaked into the ground in the gardens (and so slowly and naturally filtering into the water table) instead runs off the concrete into gutters, before entering the sewerage system through the small domestic sewers, from where it enters larger catchment sewers and then the trunk sewer network (which very quickly puts pressure on the interceptor sewers). Municipal and private temporary car parking also play a key role here, given the ever-present demand for parking in the city, as well as the dramatic expansion of the road network itself. Further, the huge expansion in industrial units, supermarkets and shopping centres (with their own large car parks), especially on the outskirts of the city, has further dramatically reduced the amount of land available for rainwater to soak into.

The result is that Bazalgette's Main Drainage system, designed from 1859 to overflow only when rainfall levels top 6mm/hr, now discharges into the River Thames through the CSOs with as little as 2mm of rainfall. Such discharges typically occur around once per week in the case of the tidal River Thames below Teddington Lock (the original Bazalgette design target catered for just twelve such discharges a year), with a mean annual 18 million tonnes of waste water being discharged into the tidal River Thames. This, of course, increases in extreme years on a like for like basis: in 2012, 25 million tonnes of waste water was discharged through the CSOs into the tidal River Thames. If we hadn't taken steps to initiate the London Tideway Improvements (including the Thames

Tideway Tunnel), then the figure could have risen to an average 70 million tonnes of waste water a year into the river by 2030.

Once in the river system, waste water immediately starts creating problems for the River Thames and those using it, not quite on the level of the 'Great Stink of London', but certainly pointing to what would happen if the problem hadn't been tackled. In Bazalgette's day the river was barren of marine life given the level of pollution, so the issue then was the danger and disruption to the human population of the city. Today, however, in the first instance, it is the wildlife that suffer. There has been a successful clean up of the River Thames in the twentieth century, with Thames Water alone spending over £3 billion on extending and improving WWTWs in the Thames Valley above Teddington Lock since privatisation in 1989. Fish have, of course, been the prime beneficiary from this clean-up, with the EA saying in 2015 that a survey it had carried out with the help of volunteers from the Zoological Society of London had identified seventeen different species in total at eight locations between Richmond and Gravesend. These included sea bass and flounder. Such reports only reveal the specific types found during a given specific survey. The tidal River Thames is a very important nursery for many fish that are commercially fished in the North Sea. Over 125 fish species live in the River Thames, throughout its length, including salmon, sturgeon, trout, catfish and eels. The danger to this generally healthy fish population is that when the waste water is discharged into the river through the CSOs during rainfall events, it absorbs the oxygen in the water, asphyxiating the fish. In a single event in June 2011, more than 100,000 fish asphyxiated in the river at Barnes in west London when 400,000 tonnes of waste water overflowed into the river, though it must be noted this was not solely related to a CSO discharge. The fish are, of course, only part of the 'faunal package' of animals living in and around the river, with all creatures, such as invertebrates that live in the river, being seriously affected. Meanwhile, 'plugs' of waste water and sewage-derived litter from the discharges can be found washed up on the banks up and down the river. In the summer, when the noxious influence is at its height given the heat, with the effects of the tide and the lower flow over Teddington Weir, it takes up to three months for this waste water to make its way downriver to Southend.

Perversely, one of the big issues regarding the initiation of the Thames Tideway Tunnel project was that when the team were trying to communicate the need to the majority of the population of the capital city who live away from the River Thames and its tributaries, this population had little visibility of these issues and therefore questioned its requirement. But the problem is there, as anyone who regularly uses the river for water sports, fishing or bird watching will know. In fact, the problem was not just there, it is growing, meaning that the water quality in the river is poor and the UK has failed to comply with the UWWTD.

Campaigners against waste water being dumped untreated into the River Thames had long been calling for action to be taken, for example Baroness Ludford of Clerkenwell, a Liberal Democrat Member of the European Parliament (MEP) for London. In 2005, she took a petition to the European Parliament to highlight the pollution problems in the river. A qualified barrister and former London councillor, Sarah Ludford was ideally placed to play a leading role in this campaign.

Her campaign was successful and led in 2009 to the UK Government being taken to the European Court of Justice by the European Commission (EC). The context here was the threat to human health and the marine environment posed by untreated waste water being dumped directly into the River Thames. In the move, the commission alleged that the UK Government was specifically in breach of the 1991 UWWTD because it had failed to treat this waste water. EU Member States had agreed to put in place adequate waste water treatment facilities to prevent this from happening in large towns by the end of 2000. Specifically, under the UWWTD Member States are required to determine the trophic or nutrient status of all of their water bodies (whether coastal, estuarine or freshwater) every four years. Assessments for this review are enacted on the basis of scientific sensitive areas' identification criteria and evidence. The waters in question must be identified as a sensitive area in the event of falling into one of the following categories: they are surface water bodies which have depleted oxygen levels or levels that are likely to fall if preventative action is not taken; they are surface fresh waters that are intended for the abstraction of drinking water and which could contain 50mg of nitrate per litre if action is not taken; or, they are areas where further treatment is necessary in order to meet other EC Directives, for example the bathing or shellfish water directives.

All of the waters draining into the catchments of the sensitive area receiving water are included in the designations under the

UWWTD. Once an area has been identified as sensitive, all quali-fying WWTWs discharging either directly or indirectly into the sensitive area must have in place, within a deadline of seven years, more stringent processes for the treatment of urban waste water.

At the time of the action, the EU's then Environment Commissioner, Stavros Dimas, said:

> More attention needs to be paid to upgrading collecting systems to ensure full compliance with EU legislation on waste water treat-ment. Such investment will bring enormous benefits in terms of improving the quality of the environment.[1]

Campaigners supporting the legal action argued that the equiv-alent of 4,000 Olympic-sized swimming pools of waste water were pumped into the River Thames between January and August 2009, taking up to three months to disperse. Baroness Ludford said at the time:

> It is scandalous that Thames pollution from sewage is continuing for so long that the Commission is obliged to take legal action. The Thames should not be used as an open sewer, with an unac-ceptable threat to our health and the environment.[2]

She was joined in her condemnation by Mike Tuffrey, then Leader of the Liberal Democrat Group on the London Assembly, who blamed the Government for not acting earlier. He said:

> The dumping of raw sewage into the Thames is something that happened in the Victorian era. It certainly should not be happen-ing in the 21st century in one of the most developed capital cities in the world. If the UK Government had taken the issue seriously and acted earlier, this action by the European Commission could have been totally avoided.[3]

Matters regarding the UWWTD came to a head on 18 October 2012 when, after a long fight, the UK was ordered by the European Court of Justice to clean up the River Thames or face multi-million-pound fines. By way of context, Greece was fined 15.9 million Euros for breaching the same directive in November 2015.[4] If the UK had been asked to pay in the same circumstances as Greece and at the same time, then its fine would have totalled some £684 million,

The effects of combined sewer overflows on the foreshore at Greenwich.

or £228 million annually (based on a maximum daily penalty of £625,000 per day).

The threat of potential breaches of the UWWTD only served to keep the UK Government focused on the issue. However, the initiative to reduce pollution in the tidal River Thames had already borne fruit much earlier, in the form of a new body set up at the turn of the millennium to determine the amount of waste water discharged into the Thames and propose viable solutions to the problem.

Thames Tideway Strategic Study Group

In 2000 the Government, under pressure from all quarters regarding the increasingly alarming levels of pollution in the River Thames, initiated the multi-agency Thames Tideway Strategic Study (TTSS) with the aim of analysing and dimensioning the scale of the problem. The study was specifically looking at how much storm waste water was being discharged into the River Thames (and the River Lee), and the physical state of the river in terms of coliform bacteria and similar indicators. After investigating, they were asked to propose solutions, while keeping costs and benefits in mind.

The TTSS comprised Thames Water, the EA, the Department for Environment, Food and Rural Affairs (DEFRA) and the Greater London Authority (GLA), with Ofwat as an observer. The independent chairman was Professor Chris Binnie, a leading specialist in dams, tidal power, water resources, water supply and flooding. One of the leading cast of characters in the TTSS and the wider Thames Tideway Tunnel story, he graduated from Cambridge with a first in engineering and then in law. He carried out his postgraduate research studying dams and hydro power at Imperial College, and in a long and wide-ranging career was Director of Water at Atkins and a panel engineer under the Reservoirs Act for thirty years. Among many other things, he was also President of the Chartered Institution of Water & Environmental Management, and was on the management board of the Tidal Power Group in the 1980s.

It was in the context of the TTSS that I first became engaged with what ultimately became the Thames Tideway Tunnel. I was Project Director in Thames Water's engineering department, and loaned a few of my project managers to the group to be the secretariat; their feedback gave me invaluable insight when the opportunity arose to eventually work on the project.

Three factors underpinned the work of the TTSS:

- Protecting the ecology of the tidal River Thames.
- Reducing the aesthetic pollution of the tidal River Thames due to sewage-derived litter.
- Protecting the health of recreational users of the tidal River Thames.

Within the TTSS process, four strategies were ultimately considered to meet the technical and environmental requirements:

- The adoption of source control and sustainable urban drainage systems (SUDS).* They are a natural approach to any requirement designed to manage drainage in and around urban developments of all types, and work by slowing down and holding back the storm flow that runs from a given site, thus allowing natural processes to break down any pollutants. SUDS techniques include the use of permeable conveyance systems such as swales, linear grass-covered depressions in the ground that lead storm water overland to a storage/discharge system, often using road verges. These are usually shallow and wide. Another candidate technique would be through the use of 'green roofs', whereby a building's roof is partially or totally covered with a waterproof membrane, on top of which vegetation is set in some kind of growing medium. Such 'green' roofs can also include added layers, like a root barrier, and also irrigation and drainage systems. Permeable pavements and soakaways also feature among candidate SUDS techniques.
- The separation of waste water and storm flow, using local storage.
- The screening, storage or treatment at discharge points into the river, including the storage and transfer tunnel idea.
- Treatment in the river itself, for example using specialist vessels to inject oxygen into the river and others to scrape the sewage-derived litter from the surface. However, this solution would only treat the symptom, not the cause.

One key factor considered by the TTSS when evaluating the above four options was that of climate change and the predicted effects it may have on rainfall intensity and patterns. To combat this, TTSS

---

* These are now referred to as 'sustainable drainage systems' (SuDS).

undertook sensitivity predictive research through to the 2080s to test the resilience of any of the solutions considered. The group devised a method by which the UK Climate Impacts Programme's (UKCIP) predictions for climate change could be used to compare the performance of any solution against any future climate change conditions. For this task, an analysis was carried out between two UKCIP rainfall predictions at Greenwich, one then (2000) and one in 2080 (they analysed the most distant forecast that might be assessed with reasonable confidence, while representing a robust assessment of rainfall).

For this assessment, the storm events predicted by the UKCIP for 2000 and 2080 were grouped into 'event size' categories, and a comparison was made between the two rainfall sets, one being divided by the other to give a ratio between the two years. The ratio was then used to calculate how the generated storm events would change between the years 2000 and 2080.

The findings, factored into the TTSS evaluations, showed that there would be an increase in greater intensity rainfall events (thus putting increased pressure on the sewerage system and increasing the likelihood of the activation of the CSOs), but also a decrease in the frequency of lesser intensity rainfall events.

The TTSS initially reported to Government in June 2004, and after considering all four strategies concluded that only one – the screening, storage or treatment at the point of discharge (featuring a 34.5km storage tunnel to intercept the overflows and convey the waste water for treatment) – would meet the objectives that they had been set. All of the remaining options were found to either be insufficient or impractical because of their inability to offer a practical solution. Because of the fifteen years it would take to build the tunnel, the Government asked the TTSS to look into smaller-scale ideas that might provide solutions at an earlier date, and also asked for additional information on the proposed tunnel option. Nevertheless, when the final report was published in February 2005 in the form of the TTSS Steering Group Report, the TTSS held firm with their initial recommendation:

The Thames Tideway Strategic study has investigated, researched and assessed the operation and environmental impact of wastewater discharges from the collecting system and treatment works on the river. Objectives and possible solutions have been developed which have been subject to cost benefit analysis.

This work indicates that parts of the London collecting and treatment system require improvement to meet one or more of the objectives.

The study has established that the environmental objectives can only be fully met at least-cost by completing both the quality improvements to the treatment works discharges and by provision of a storage-and-transfer tunnel.

It is for Government to decide whether the preferred option identified by the Study proceeds and at what pace. The Steering Group has received a request for additional investigations to be carried out to inform this decision and to consider smaller scale measures that could bring earlier improvements to the Tideway. This work is under way and will be reported on during 2005.

An outline delivery timetable for the storage-and-transfer tunnel has been developed and confirms that a five-year period of detailed engineering design and planning would be required. Construction could take a further 8–10 years, so overall solution delivery within 15 years is believed feasible.

A supplementary report to Government in November 2005 emphasised the same conclusions. At that point Ofwat, who had a regulatory responsibility to make sure Thames Water continued to be financeable, as well as looking after the interests of customers who might suffer, tasked Jacobs Babtie Engineering Group to look at alternatives to the TTSS recommendation. Their report was finally published in February 2006, recommending alternatives to the TTSS's recommendations. Slightly earlier, in December 2005, a DEFRA-led working group was established to examine both of the reports and compare. Subsequently, in a 27 July 2006 letter to Thames Water from DEFRA's Minister of State for Climate Change and Environment, Ian Pearson, the Government told the company to develop two alternative tunnel options, the work to be completed with the support of the EA and Ofwat by the end of 2006. The options to be considered were:[5]

Option 1: A tunnel more than 30km in length designed to intercept intermittent discharges from unsatisfactory overflows along the length of the tidal River Thames, and then to convey the waste water for treatment in east London. The specific alternatives within this option are detailed below:

Main Option 1a: the complete main tunnel of 7.2m diameter running from Hammersmith to Beckton with a link tunnel capturing Abbey Mills waste water discharges that joined the main tunnel at Charlton. This option required additional treatment capacity at Beckton of 536ml per day (Ml/d), giving a total flow to full treatment (FFT) at Beckton of 2336Ml/d (when additional upgrades were included), this additional treatment being required to allow the tunnel to be emptied and the flow treated in forty-eight hours after a storm had receded. This was TTSS's preferred option.

Variant Option 1b: the complete main tunnel but this time of 6m diameter. Option 1b was a variant of Main Option 1a, with a tunnel route and set-up being similar, but a smaller tunnel diameter of 6m, and hence a smaller storage capacity. As the storage was smaller than 1a, Option 1b required a smaller additional treatment capacity at Beckton of only 305Ml/d, giving a total FFT at Beckton of 2105Ml/d (including additional upgrades).

Variant Option 1c: again featuring the complete tunnel (with direct Abbey Mills link). Option 1c had the same volume and dimensions as Option 1a, but with a direct tunnel linking Abbey Mills to Beckton (later called the Lee Tunnel) as opposed to it simply joining the main tunnel at Charlton.

Option 2: Two shorter tunnels, in west and east London, to intercept intermittent problem discharges along these stretches of the river, and probably additional treatment in east London. This solution was based on the Jacobs-Babtie alternative recommendations following their engagement by Ofwat.

Main Option 2a: an east–west tunnel with a direct Abbey Mills link some 13m in diameter. This was a two-part tunnel solution with a western tunnel linking Hammersmith to Heathwall and an eastern tunnel linking Abbey Mills to Beckton STW. The CSOs in between Heathwall and Beckton were not to be intercepted under this concept, where the western tunnel returned stored discharges to the sewerage system once a storm had subsided, it eventually flowing back to the Beckton and Crossness WWTWs for full treatment. This option required additional treatment capacity at Beckton of 198Ml/d, giving a total FFT there of 1998Ml/d (when additional upgrades were included). This additional treatment was required to allow the eastern and western tunnels to be emptied and treated in forty-eight hours after the storm.

Variant Option 2b: an east–west tunnel with a direct Abbey Mills link, 10m in diameter. Option 2b was identical to Option 2a, except the eastern section of the tunnel had a smaller 10m diameter and hence a smaller storage capacity. This variant solution was designed to have a similar residual spill frequency to Option 2a but because it had a smaller storage capacity it required a much larger increase in treatment capacity at Beckton to ensure that the tunnel could be pumped out to treatment whilst the tunnel was still filling. This option therefore required additional treatment capacity at Beckton of 900Ml/d, giving a total FFT at Beckton of 2,700Ml/d (when additional upgrades were included).

Variant Option 2c: an east–west tunnel from Abbey Mills to Beckton link via Charlton. This option was similar to Option 2a with the exception that the eastern section of the tunnel diverted to, and intercepted, the Charlton CSO before terminating at Beckton. The storage capacity was similar to that of the eastern section of Option 2a, requiring the same treatment set-up with an additional treatment capacity at Beckton of 198Ml/d giving a total FFT at Beckton of 1998Ml/d (when additional upgrades were included).

London Tideway Improvements

Thames Water reported back to Government on schedule in December 2006, recommending what became known as the London Tideway Improvements. This featured a three-pronged engineering solution based on Option 1c, comprising improvements to existing infrastructure (WWTWs), a new tunnel to improve the River Lee and, finally, the Thames Tideway Tunnel. To start, Thames Water had to upgrade the five existing tidal waste water treatment works servicing London:

- Mogden treatment works, Isleworth, had a £140 million upgrade to increase its treatment capacity by 50 per cent.
- Crossness treatment works, London Borough of Bexley, had a £220 million upgrade to increase its treatment capacity by 44 per cent.
- Beckton treatment works, London Borough of Newham, had a £190 million upgrade to increase its treatment capacity by 60 per cent.
- Riverside treatment works, Rainham, had a £85 million upgrade as part of a major scheme to help improve the water quality in the River Thames, and to also produce renewable energy.

- Long Reach treatment works, Dartford, had a £40 million upgrade.

This phase of work was completed by Thames Water in 2014, with the total costs being £675 million.

The next step was to construct the deep storage and conveyance Lee Tunnel. This was designed to run for 6.9km from Abbey Mills to Beckton at depths of up to 75m, the tunnel being designed to capture 16 million cubic metres of waste water per year from the single largest polluting CSO in London. In January 2010, the work on the Lee Tunnel was contracted out by Thames Water to a joint venture called 'MVB' comprising Morgan Sindall, Vinci Construction Grands Projets and Bachy Soletanche, construction beginning in that year. The Lee Tunnel was completed in 2015 at a cost of £635 million, and was opened by the then Mayor of London Boris Johnson on 28 January 2016. It was the largest project ever constructed in the UK privatised water industry at that time.

The final step was to build the Thames Tideway Tunnel. The river modelling and planning for this programme was facilitated in association with the EA, creating a catchment model featuring all of the sewers in the catchment to determine the storm flow and waste water inputs. The most important decision here was to decide how many CSOs would be intercepted. In the end, thirty-six of the fifty-seven worst-polluting CSOs were selected, and then categorised by the volume of their flow and discharge per year. Of these thirty-six, that at Wick Lane was dealt with under a separate project by Thames Water, while the CSO at Abbey Mills was already provided for by the Lee Tunnel above. That left the remaining thirty-four CSOs to be catered for by the Thames Tideway Tunnel, the other twenty-one CSOs not being polluting enough to incorporate (some of them only sporadically releasing waste water while others had been bricked up).

And thus was born the Thames Tideway Tunnel, the means by which 95 per cent of the flow from the thirty-four CSOs will be managed, together with 90 per cent of the 10,000 tonnes of sewage-derived litter removed, once the tunnel is in operation in early 2024. Although the Thames Tideway Tunnel needs to manage the flows from thirty-four CSOs, subsequent design showed that not all thirty-four CSOs needed to be intercepted for their flows to be adequately controlled.

4

RISING TO THE CHALLENGE

The pathway that led Thames Water to focus on large-scale storage and transfer tunnels (the London Tideway Tunnels comprising the Lee Tunnel and Thames Tideway Tunnel), together with supporting infrastructure, as the solution to London's sewerage system requirements was long and winding. By December 2006, the Government had Thames Water's recommendations in front of them and things began to gather apace.

I cannot emphasise enough how bold a decision this was on the part of all the stakeholders, from the TTSS process onwards, to build the Thames Tideway Tunnel as well as the Lee Tunnel. The Thames Tideway Tunnel will ultimately be a 25km long, 7.2m diameter tunnel at depths of up to 65m (30m at the Acton Storm Tanks in the west through to 65m at the Abbey Mills Pumping Station in the east) mainly beneath the River Thames (giving it a self-cleansing gradient falling from west to east), all aimed at intercepting overflows from Bazalgette's original interceptor sewers. This is where I join the story, initially as Head of London Tideway Tunnels and then of Thames Tideway Tunnel.

The Programme Begins and Implementation Strategy

Arguably, the single most important event in the whole story of the Thames Tideway Tunnel occurred on 22 March 2007 when the Government announced that a decision had been made to back the development and implementation of the full-length storage and transfer tunnel solution based on tunnel Option 1c, following receipt of Thames Water's December 2006 report and a subsequent

Regulatory Impact Assessment (published on the same date as the Government announcement). Things then really began to move forward on 17 April, when Ian Pearson, still DEFRA's Minister of State for Climate Change & Environment, wrote to David Owens (then CEO of Thames Water) asking the company to progress the full-length storage and transfer tunnel, together with the Lee Tunnel and additional infrastructure upgrade and construction recommendations (the London Tideway Improvements):

> My view is that early phasing of the Abbey Mills to Beckton Tunnel, as well as work on the rest of the scheme, will be needed in order to make progress toward compliance with the (EU) Directive (and associated duties under the Water Industry Act 1991 and the Urban Waste Water Treatment (England and Wales) Regulations 1994 ('the 1994 Regulations')) as quickly as possible. This is because early phasing of this tunnel would enable 50% of the total volume of collecting system overflow discharges (those to the River Lee from Abbey Mills Pumping Station) to be addressed well before completion and the long tunnel.
>
> I am writing to request that Thames Water makes provision for the design, construction and maintenance of a scheme for the collecting systems connected to Beckton and Crossness sewage treatment works which:
>
> Involves a full-length storage tunnel with additional secondary treatment at Beckton sewage treatment works;
>
> Meets the requirements of the 1994 Regulations, including for sewerage undertakers to ensure that the design, construction and maintenance of collecting systems is undertaken in accordance with best technical knowledge not entailing excessive costs (BTKNEEC);
>
> Complies with such discharge consent conditions as will be set by the Environment Agency, in exercise of its duty under regulation 6(2) of the 1994 Regulations, to secure the limitation of the pollution of the tidal Thames and River Lee due to storm water overflows; and
>
> Limits overflow discharges at Abbey Mills Pumping Station as soon as possible.[*]

---

[*] For a look at the full letter, please see Appendix A.

After the Minister had asked Thames Water to progress with an Option 1c solution, the scene was set to begin the Thames Tideway Tunnel (called Thames Tunnel at this time) project, led by Thames Water and supported by the EA and Ofwat. Although work had progressed in enough detail to provide an initial estimate of cost, further work was required to develop this solution to a stage where a planning application could ultimately be prepared and submitted, and designs could be progressed to a sufficient level of detail to allow tendering for the work implementation to take place.

Initial Planning: Priorities

From the outset, with DEFRA telling Thames Water to go ahead with Option 1c and the company being set the task of taking forward the London Tideway Improvements, the strategy was to go ahead with the programme as soon as possible, and therefore to achieve the consequent benefits at the earliest practicable date. Thames Water never lost sight, however, of the need to carry out this work in a way that provided value for money, met planning and environmental requirements, and was sustainable in terms of financial and future demand.

It was decided early in the process that tackling the enhancements for the five WWTWs and building the Lee Tunnel were priorities over the building of the Thames Tideway Tunnel. This was primarily because the associated Abbey Mills Pumping Station CSO discharges account for 40 per cent of the total spill volume from all unsatisfactory CSOs, but it was also more straightforward to develop, from both a planning and an engineering perspective. As detailed in Chapter 3, the WWTW upgrade programme was completed in 2014 and the Lee Tunnel in 2015, the latter being opened in January 2016.

The next focus was the Thames Tideway Tunnel, considerably more complex than either the WWTW upgrades or the Lee Tunnel, both in terms of engineering (it being one of the most complex engineering tasks ever undertaken in the country) and also the planning/interface with key stakeholders and other interested parties. The programme also required large sites to be acquired that were not at that time in Thames Water's ownership.

Boat tour briefing of key stakeholders in July 2009, including Government representatives and senior officials from riparian London local authorities.

Initial Planning: Programme Management

Thames Water planned from the outset to provide the client team for the development and implementation of the Thames Tideway Tunnel project, with the plan being to engage relevant programme managers to ensure timely, on cost and sustainable delivery of the project. The aim here was to make sure the outline scheme was far enough advanced that it could be fully developed and costed well before the consultation and planning application began, this being anticipated for around 2010. At this point Thames Water focused on:

- Considering all aspects of the design, including the main tunnel, connection tunnels, access shafts, CSO interceptions and all of the above-ground works.
- Getting a better understanding of the ground conditions needed to minimise risk and to determine the appropriate tunnelling technology to be employed for this complex project.
- Confirming the route of the tunnel and connection tunnels with at that time further design development work being needed, particularly with regard to the location of tunnel main drive and

reception shaft sites and CSO inception sites. Some sites were in Thames Water's ownership but many were not.

- Determining the logistics and practicalities of constructing and operating out of the sites along the route so that, through the consultation and planning process, they could be understood and agreed with stakeholders.
- Putting in place plans for the re-use of the excavated materials from the construction process, including any required infrastructure support proposals. From this early stage the use of the river was the preferred option here, where appropriate, but not all of the proposed sites were adjacent to the Thames.

Initial Planning: Design

From the outset, the design itself was developed to ensure its constructability (a project management technique that reviews all of the construction processes of a project from start to finish during pre-construction phase), with early involvement of specialists and contractors to facilitate this process. This initial phase of design development, which started in 2008, was key to providing supporting information for the vital planning application. The intention was to take the design work forward up to some 20 per cent of the total design before tendering began for the construction contracts.

Initial Planning: Cost Estimates

After this initial work was carried out it was estimated that the total cost of the project would be in the region of £3.1 billion. Under the business model, this would be paid for by Thames Water customers through their waste water bills, hence the ongoing interest of Ofwat.

Initial Planning: Consultation and Planning Application

The aim was to submit this as early as possible to give it the best chance of being approved quickly and for construction to begin. The preliminary studies work had already been completed by that point through the TTSS and subsequent analyses, as Thames Water had decided early on to separate out the planning applications for the component parts of the London Tideway Improvements. All parties involved acknowledged that, while planning (if not construction) work on the Thames Tideway Tunnel would commence at the same time as the other components of the London Tideway Improvements, the former was far more complex and would need to take account of a longer programme to assess and acquire the sites not then currently in the ownership of Thames Water. Furthermore, the programme had already built into its timeline a protracted planning process and the likely requirement for a Public Inquiry before the key planning decision was finalised.

Initial Planning: Construction

It was clear from the very beginning of the project that it would be highly complex with regard to procurement and construction, with significant technical, commercial, planning and environmental risks evident. It was considered fundamental to the successful delivery of the project to work closely with specialist construction contractors, tunnelling machine and lining suppliers, as well as construction experts at the very earliest stage of project development in order to proactively mitigate the risks and develop optimal enabling works and temporary/permanent works proposals. All of these fed into the consultation and planning application process. It was also set in stone that the vast majority of work would be competitively bid, with an adequate amount of design work being carried out in advance of any tender award to reduce risk and provide cost certainty (while accepting at the time that the tunnelling work would always have a degree of retained risk due to the very nature of the work involved).

Initial Planning: Finance and Funding

The risk profile for the construction of the Thames Tideway Tunnel was fundamentally different from any that Thames Water had undertaken to that time, and such risk could not be mitigated through the company's day-to-day capital programme. It was therefore essential that the existing regulatory incentives and recourse mechanisms should be developed at an early stage to better define and allocate the risks, and also to ensure that the overall returns were consistent with the business risk that the capital markets,* through Thames Water, were asked to assume. This would enable both Thames Water and Ofwat (looking after the interests of Thames Water customers) to fulfil their legal obligations, such that Thames Water continued to be able to finance its functions. Thames Water accepted that the characteristics of such

---

* A capital market is part of a financial system concerned with raising capital by dealing in shares, bonds, and other long-term investments.

large projects and their risks were such that alternative delivery and financing models would also have to be considered in order to ensure best value for customers, and therefore a number of alternative funding options were explored with Ofwat. The initial task here was to agree a suitable mechanism with Ofwat to fund design development work, which would then enable the costing to be advanced to test the funding options against the market. The early completion of market testing would then help inform the decision on the most appropriate funding method going forward for the project. The selected funding solution would have the support of the capital markets.

Initial Planning: Programme and Key Milestones

Thames Water, the EA and Ofwat aimed to implement the project to a timescale that was efficient, but also – crucially – the best value for money for the business and Thames Water's customers (a recurring theme, as the reader can see). The key milestones at the beginning of the programme are detailed below. They assume that the question of funding will have been resolved by the dates shown, and also that planning permissions would have been obtained within the timeframe.

June 2007: Agreement on funding for development phase.
September 2007: Agreement on principles of funding with Ofwat.
May 2008: Planning application submitted for the Lee Tunnel.
September 2011: Planning application submitted for the Thames Tideway Tunnel.
July 2012: Work completed on the Lee Tunnel.
May 2020: Work Completed on the Thames Tideway Tunnel.

The date for completing the second phase of the London Tideway Improvements – the Lee Tunnel, from Abbey Mills Pumping Station to Beckton WWTW – had been carefully reviewed early on, taking account of the delivery route that incorporated an appropriate level of design and planning work before award of the main construction works. This was to enable the procurement process to take place effectively and on a competitive basis.

Initial Planning: Regulatory Considerations

Thames Water recognised at the beginning of the programme that it would not be possible to move the project forward without working closely with the EA and Ofwat. This was because the Thames Tideway Tunnel was acknowledged to be a complex integrated solution to the challenge of updating a Victorian-era sewerage system. Therefore, the three organisations worked together collaboratively on the project from the very beginning.

The Project Matures

In March 2007, the Government gave final approval to start work on the Thames Tideway Tunnel, based on all of the research, analysis, planning and examination of alternatives to that date.

At this point the route of the proposed tunnel was Acton to Hammersmith, both deep underground and under land, and then following the route of the River Thames to Beckton WWTW, the overall length being 32km. Under this signed-off plan it would sit between 30m and 70m below the ground and have a diameter of 7.2m. Along its course it would manage the flows of thirty-four of the CSOs that connected the existing sewerage system to the river. The interception of these CSOs allowed for the overflow of waste water and surface water to be diverted into the tunnel for storage, and then onward conveyance to the Beckton WWTW and subsequent discharge as treated effluent.

The programme accelerated from that point. After a competitive tendering process, CH2M Hill (part of Jacobs since 2017) were appointed as the programme managers for the Thames Tideway and Lee Tunnels in March 2008. CH2M Hill were a global engineering company providing programme management design, consulting, construction and operations services for corporations and government at national and local level, who were proud enough of their Thames Tideway Tunnel work to publicise it on their website under the title 'Transforming the Ecology of London's River'. At the time of the contract award, Steve Walker, Thames Water's Major Projects Director (who had 'owned' the project from 2006 and was a key driver), said:

The (Thames) Tideway Tunnel scheme is Thames Water's biggest single investment project by far. These exceptional tunnels will have enough capacity to store millions of litres of diluted sewage and transfer it to our Beckton sewage treatment works. This scheme is essential if we are to improve the quality of the river and reduce the environmental impact of sewage overflows. To successfully implement this massive and challenging engineering project we need the full support of our stakeholders and a world-class team with the appropriate knowledge and experience to take it from planning to construction. We are delighted to have CH2M Hill on board to help us deliver a sewer system fit for the 21st Century and beyond.

Lee McIntire, then president and chief operations officer for CH2M Hill, responded:

We look forward to working with the Thames Water team to deliver this landmark engineering programme. Together we will help the city meet current and future waste water requirements, and leave a lasting legacy that will improve the quality of the Rivers Thames and Lee for future generations.

Meanwhile, I was recruited and began my new role on 1 April 2008, a position I held until the official formation of Bazalgette Tunnel Ltd on 24 August 2015. A key part of the first set of activities was for me to set up the project team to deliver both of the tunnels. This was initially based in Ambassador House in Richmond, with a chartering session being run by CH2M with the Thames Water staff at the start of the project. I had three people report to me at this time: Sian Thomas, project manager for the Thames Tideway Tunnel; Nick Butler, project manager for the Lee Tunnel; and John Greenwood, design manager. The original CH2M programme manager was Mike Dominica, who took the programme through the mobilisation phase and then passed the mantle to Jim Otta. Jim went on to make a massive contribution to the project. One of the first documents we produced was the vitally important Site Selection Methodology that was signed off by all the relevant local authorities. Our initial work at this time was focused on route selection and the development of outline design, which included tunnel drive strategy and site selection. This led to the shortening of the route from 32km down to 25km, made possible by lowering the Lee Tunnel at Abbey Mills Pumping Station. This latter innovation saved some £700 million of project cost and did away with the deepest section of tunnel.

We moved from Richmond to Paddington in November 2008, and this proved to be absolutely key because this office provided far more suitable accommodation, was in the catchment of the project and also had good accessibility to Reading (where Thames Water is headquartered) and the rest of London. Various office options were investigated, including the old Woolworth's building on the Marylebone Road, but The Point in Paddington Basin proved to be a great choice and until recently was still the main base for the project. Our task at both Richmond and Paddington was to build a 'virtual team' around CH2M, augmented with staff from the many companies in the Thames Water Framework Agreements. The great success of this was – and still is – that in the project team you could not tell who worked for whom. We have always ensured that we have an energised and motivated workforce.

With the 2008 appointment of CH2M, another key figure had already entered the stage of the Thames Tideway Tunnel story, a true partner in crime for me as our programme progressed. This was Nick Tennant, who was appointed communications manager under CH2M for Thames Tideway Tunnel. Nick had previously been head of communications for Thames Water, working with Richard Aylard who, as will be seen, is another key figure in the story of the Thames Tideway Tunnel.

From 2008, a great deal of work by the team went into considering the best way forward to obtain planning permission. All the various options were considered, including Hybrid Bill

Opening of Sir Joe's Cafe at The Point in March 2014. Cupcakes for all!

(a set of proposals for introducing new laws, generally used for major infrastructure projects), Development Consent Order (a means to obtain permission for Nationally Significant Infrastructure Projects) and making fourteen separate applications (under the Town and Country Planning Act 1990) to relevant local authorities. Given DEFRA's reluctance to pursue the Hybrid Bill route, the Development Consent Order route was preferred.

Meanwhile, the next project milestone was when Steve Walker left the project in April 2009, being in the first instance replaced by David O'Reilly, who then left six months later, leaving me to report directly to David Owens, the Thames Water CEO. These were challenging times for me as, despite the project ramping up dramatically, I was asked to significantly reduce costs. In order to do so, the workforce numbers were reduced from seventy-two to forty overnight in November 2009. Owens himself left Thames Water in December 2009, after which I then reported to the new Thames Water CEO Martin Baggs, a chartered civil engineer who had a good appreciation of large projects such as ours.

The team ramped up again in April 2010, to prepare for the first phase of the consultation, which was planned for June, though this was delayed to September due to the general election of that year and the subsequent creation of the Coalition Government.

By this time the estimate for the project was £3.6 billion* (though following the consultation process detailed in the following chapter, and subsequent changes, this was increased to £4.2 billion,** a level it has remained at since). Having been responsible for obtaining planning permission and procurement of main works contractors for the Lee Tunnel, after contract award the project was handed over to Lawrence Gosden in early 2010, the then Head of Capital Delivery at Thames Water.

My next task was to initiate the preparation of the documents that would be needed for the start of the Phase One consultation, such as the Needs Case and the System Master Plan.

The team continued to grow, with Mike Gerrard joining us in May 2011 as managing director of the Thames Tideway Tunnel. Mike is a chartered engineer and has vast experience of financing major infrastructure projects. At that time, Steve Walker, Jim Otta and I were on a tour of Southeast Asia and China to raise market

---

* in 2009 prices.

** in 2011 prices.

awareness of the project. Dave Wardle, the London Area Manager for the EA, began to make his supportive presence felt from this point onwards too, and indeed has been a fantastic stalwart for the tunnel project ever since (there have been two embedded EA members within the project team since April 2008). Thus, by 2011 all of the key senior team figures were in place to progress our project. They formed what I liked to think of as our premier league big hitters!

That brings me on to the wide-ranging stakeholder consultation for the project. We recognised early on in the programme that it was important to ensure key stakeholders were fully engaged from the design phase onwards, so that the final proposals and subsequent construction were fully supported and endorsed, and to make sure that any issues that arose were quickly dealt with. It has always been key for us to get out to speak to stakeholders first-hand, with thousands of hours being spent in the evenings engaging with communities and residents. I believe that we won their trust by being honest, open and transparent, this engagement being a new and vital competence for the 'modern engineer', with face-to-face meetings being very important. Further, I found it

Talking to Rita Cruise O'Brien, Chair of SaveYourRiverside, at the Chambers Wharf exhibition in November 2011.

was critical to the success of our project to be empathetic and sympathetic with stakeholders. That is why, throughout our project, there have been over 400 external presentations on the proposed work, with me doing about 250. The project team has never refused to give anyone a briefing. In addition, there were 114 days of public exhibitions across the fourteen local authorities during the main phases of consultation. One key and specific stakeholder group deserves particular recognition here: the London public most directly affected by the Thames Tideway Tunnel. The construction activities, as of the autumn of 2017 (both in terms of main tunnelling sites and the many CSO interception sites), are very visible to these local residents. To mitigate the impact of this, we attempted to develop a close understanding with local communities early on in the consultation process, so that we could discuss the benefits and impacts of the project, make sure that there were no surprises and ensure expectations were managed. This was sometimes successful, but at other times it was more challenging. Despite our best efforts, it is clear that we will not make everyone happy all of the time.

All of this external engagement activity was resourced by the project team members themselves, it being a great experience for young professionals. All of those involved in this engagement were thoroughly briefed before each session and given the latest up-to-date site-specific information, with particular reference to who might attend and their views on the project.

Construction Works Procurement Strategy

Both the construction works and Infrastructure Provider procurement were the responsibility of Amar Qureshi, the commercial director. The procurement team achieved a remarkable feat of nearly forty procurements (of which the main works contracts count as only one procurement) without a single successful challenge. Very rare in the modern world.

During 2012, work was progressing on the Construction Works Procurement Strategy, this being finalised in March 2013. It included three discrete categories, these being the main works (to be delivered by the Infrastructure Provider), the Thames Water enabling works and the Thames Water system works. The strategy did not include the procurement of the Infrastructure Provider (this being considered in its own section below). The main works included the main tunnel, all connection tunnels, associated tunnel drive and reception shafts and CSO interceptions with associated drop shafts.

The Thames Water enabling works comprised a variety of contracts, ranging from archaeological investigations, demolition, bathymetry, marine works, environmental baseline monitoring and utility diversions to temporary power supplies that would be required for the tunnel-boring machines. The Thames Water system works included managing some of the CSO discharges by locally modifying the existing sewerage network.

Here I focus on the main works. The procurement processes were compliant with all European procurement legislation and UK Government Regulations. Each stage of the procurement process was carried out fully in accordance with Thames Tideway Tunnel team's procurement management processes. The approach was thorough, robust and treated all parties fairly. In addition, a 'red review' was undertaken by independent industry experts in major infrastructure procurement to challenge and test the general approach and ensure lessons learned from other major programmes were utilised.

There was a massive market engagement exercise over the five years prior to the tender documents being issued. I attended dozens of conferences over this period to inform the market of progress on the project and our strategy for delivery, together with anticipated milestone dates. At one such event, at a British Tunnelling Society presentation at the Institution of Civil Engineers in Westminster, the Telford Lecture Theatre was full and there was a live CCTV link to another lecture room in the same building. It was estimated that 350 people were in attendance. In the same period, I also had scores of separate meetings with many senior representatives from various major contractors and consultants. I knew many of these people from my previous engineering roles in Thames Water.

Prior to the production of the procurement strategy, an extensive market testing exercise was undertaken to ascertain the market appetite for the project and to understand the market's view on a number of delivery issues regarding various procurement and packaging options. The results can be summarised as follows:

- 89 per cent of respondents expressed a preference for three separate main works packages.
- 72 per cent of respondents stated that they would bid for all three packages, even if they were only allowed to win one of them.
- 78 per cent of respondents expressed a preference for the NEC3 Option C (target cost) form of contract.

The Contract Packaging Strategy was informed by market analysis and supplier feedback. The strategy settled upon comprised a main works package – West, Central and East – with marine logistics and transportation included, as well as a systems integrator package. It was clear from the very early days that the tunnelling would be split into three sections because of London's geology and the need to have shafts where the geology changed at the tunnelling horizon. The split also made sense in terms of market capacity. In addition, initially plans were for separate procurement for mechanical, electrical, instrumentation, control and automation (MEICA). This was later changed and MEICA was included in the main works packages. The three main works packages used the NEC3 Option C (target cost) form of contract and the system integrator package used the NEC3 Option E (cost reimbursable) form of contract.

The western section (including seven sites located between Acton Storm Tanks and Carnwath Road Riverside) had an initial construction value of £416 million and was approximately 6,950m long, tunnelling through London Clay. The central section (including eight sites located between Falconbrook Pumping Station and Blackfriars Bridge Foreshore) was divided into two tunnel drives: a westerly one of approximately 5,000m, and an easterly one of approximately 7,680m that reached Chambers Wharf. These tunnels will go through London Clay, the Lambeth Group (laterally and vertically bedded sands, gravels, silts and clays) and the sedimentary strata of the Thanet beds. The contract value for the central section was an initial £746 million. Finally, the eastern section (including six sites located between Chambers Wharf and Abbey Mills Pumping Station) was approximately 5,530m long, tunnelling through chalk and with an initial contract value of £605 million.

It was decided that the tenderers could bid for any number of packages but could only win one, unless in exceptional circumstances. In addition, design and construction contracts would be let for the main works packages to encourage the optimum solution to be delivered. The Crossrail experience had demonstrated that separate design and construction contracts were problematic, for a number of reasons, the main one being the lack of contractor input into the design and consequential rework of the design once construction contracts were left.

Industry Day

A great deal of effort went into the Industry Day in summer 2013. The purpose of the day was to engage with the supply chain and to inform them about the project and the timescale for the procurement activities, i.e. when tenders would be issued. It was a crucial activity to ensure maximum interest in contractors wanting to bid for the main work construction packages. It combined various presentations, Q&A sessions and a boat trip to see a number of the proposed sites. It was very professionally managed and well attended. Jim Otta compered and introduced me as the project's own 'Olympic hero' – quite an introduction. The feedback on the day was exceptional; some contractors and suppliers, who had worked in the industry for over thirty years, described it as the best Industry Day that they had ever attended.

The Official Journal of the European Union (OJEU) Contract Notice followed, together with the pre-qualification process and invitation to tender. The project was successful in achieving a competitive field of joint ventures to bid for each of the three main works packages. For the reader's background, the OJEU is a means by which all public sector tendering and contract opportunities within the EU are published.

The invitation to tenderers (ITT) was issued to four tenderers for the West main works contract on 10 December 2013. The ITT is the initial step in competitive tendering, in which supplier and contractors are invited to provide offers for supply or service contracts. The documents were returned on 21 May 2014. The East's ITT was issued to five tenderers on 14 January 2014. Tenders were returned as expected on 17 June 2014. Finally, for Central the respective ITT issue and return dates for four tenderers were 22 April 2014 and 23 September 2014.

I was part of the extensive team of tender assessors. The process was exemplary and better than I had ever witnessed. The sections the assessors were reviewing were only available in a supervised quiet room. We could only view our specific sections and we had to mark and give our comments in the controlled environment.

Indeed, the experience was like taking an exam. At least two individuals assessed each section of the tenders and the scores were moderated by a third person. It was a very robust process. Following the process three contracts were awarded for each of the separate geographical areas on 24 August 2015.

A similar robust process was undertaken to award the system integrator contract.

Infrastructure Provider

There were three key dates that played a major role in the development of the Thames Tideway Tunnel prior to the appointment of the Infrastructure Provider.

8 April 2010
The first (going backwards in time slightly) was the advent of the Flood and Water Management Act 2010. This amended the Water Industry Act 1991 by inserting a new Part 2A that conferred powers onto the Secretary of State for Environment, Food and Rural Affairs to make regulations about the provision of infrastructure for the use of water undertakers or sewerage undertakers (such as Thames Water). This set the legislative framework for the Thames Tideway Tunnel project.

12 September 2014
The second date was the date the Government announced the crucial granting of the DCO for the Thames Tideway Tunnel to be built and operated. A vital piece of legislation for the project, a DCO is the means by which permission for developments categorised as Nationally Significant Infrastructure Projects (NSIP) are obtained, including those relating to energy, transport, water and (in this case) waste programmes. I believe DEFRA opted to go down the DCO route rather than another means of facilitating the programme's approval because the department had never managed a project on this scale before. Having chosen this route, DEFRA then established a Waste Water National Policy Statement that set the framework against which the whole programme would be taken forward. This was not wholly focused on the London Tideway Improvements, but it was with regard to our activity with the River Thames that it had its biggest impact.

As an interesting aside here, this statement was itself open to a consultation exercise that many of those who opposed our tunnel solution missed out on, to their later regret, given that it set in place the national policy regarding waste water provision (again, with the London Tideway Improvements as a key component).

The date on which the DCO was granted was also a very important date for me personally, as it marked the moment at which the real 'point of no return' was reached, after which I truly believed the programme would go forward to completion. Reflecting the complexity of the project, there were two Government representatives granting and signing the DCO, in the form of then DEFRA Secretary of State Liz Truss MP and Communities and Local Government (DCLG) Secretary of State Eric Pickles MP. Making the announcement, the Government specifically said that the 'Secretaries of State conclude that on balance there is a good case for making an Order granting development consent for the proposed development', the decision being based on the Planning Act 2008. At the time, Truss said:

> In the 21st century, London should not have a river that is polluted by sewage every time there is heavy rainfall … the Thames (Tideway) Tunnel is considered to be the best solution to address London's outdated sewerage infrastructure.

Pickles added:

> This is a challenging infrastructure project, but it is clear that the Thames (Tideway) Tunnel will help modernise London's ageing Victorian sewerage system, and make the River Thames cleaner and safer.

13 July 2015

The third key date was just under a year later, when the newly created, special purpose company Bazalgette Tunnel Limited (as detailed above, called Tideway by way of trading name) was appointed the preferred bidder as the Infrastructure Provider (IP, under the above Act) to finance and deliver the Thames Tideway Tunnel Project.

Appointment of the Infrastructure Provider

The story regarding Tideway's appointment as the IP actually began earlier, with Thames Water's business case to allow the IP procurement process to commence being a key feature of its original Outline Business Case (OBC), developed between late 2011 and July 2014. Thames Water had a statutory duty, under guidance from the Government, to run the procurement as a tender, and so were not able to award the IP contract directly to Tideway. Both the Government and Ofwat had their own specific reasons to pay close attention to this process, in the case of the former because of the Government Support Package (GSP) they awarded to help enable the whole IP procurement process, and for the latter the fact that it would be the awarder of the IP Project Licence. The GSP was provided by the Government to address extremely low probability scenarios that were beyond the capacity of the private sector to absorb on terms that offered value for money to Thames Water customers. Therefore, tight attention to detail and high levels of due diligence were a core aspect of the way Thames Water ran the whole IP procurement process. It should be noted at this point that Ofwat had earlier held a consultation on the regulatory framework that was to govern the selected IP, this taking place in October 2014 on the draft project licence and other regulatory documents relevant for the IP. The consultation was deemed to have been successful.

Moving on, it was also clear from an early stage that the three parties (Thames Water, the Government and Ofwat) needed to liaise closely and clearly understand their respective roles and processes, given the opportunities for misunderstandings or miscommunications that could derail the process due to the complexity of the programme.[1] In particular, it was considered essential that Thames Water should work closely with the Government to develop suitable processes to ensure the selection of a winning bidder in full accordance with the relevant, very tight procurement laws, based on the agreed GSP, with the ultimate result being acceptable to the Government of the day.

Under procurement rules and the Utilities Contracts Regulations 2006, Thames Water was now required to tender for the procurement of an IP. The development of the strategy for this involved a detailed analysis of how the process would best be designed to reflect the nature of what was being procured; namely, the provision of finance ownership and governance for an operational business featuring all of the relevant contractors in place on time, and with the necessary commercial, technical and management capability already on board.[2]

This included extensive consideration of the extent to which the procurement process could be managed along the lines of a typical Mergers and Acquisitions (M&A) process, or whether it instead should be managed in line with a public infrastructure project procurement project. This reflected advice from project financial advisors UBS (appointed in 2012) that many of the potential bidders might not be familiar with public procurements, and also that the project might attract greater interest if it were structured and run in way that was akin to an M&A process. This process should be as simple as possible, with the key aspects being: the facilitation of the vendor due diligence; the swift and accurate response to bidder questions and requests for additional information; the imposition of a limited administrative burden on bidders; a focus on the sale price rather than other bidding requirements; and a short timescale overall, given the time constraints envisaged in building the Thames Tideway Tunnel.

The conclusion reached by the procurement process stakeholders was that in order to attract the interest of the widest possible spread of investor groups to tender for the IP procurement, the overall process would be aligned as far as possible with processes usually used for M&A transactions. It was also acknowledged at the time that the IP procurement rules would need to be adhered to tightly, given the decision to go down this route rather than the more traditional public procurement route, and also that the timetable would be longer than normal for an M&A (partly because of the need to involve all stakeholders at key stages in the process). However, the overall structure was specifically designed to be immediately recognisable to participants familiar with an M&A process.[3]

Thames Water issued the OJEU Notice and the Pre-Qualification Questionnaire (PQQ) letter on 10 June 2014, commencing the PQQ phase regarding the IP procurement. The PQQ process was designed to allow all types of bidders to express an interest in fulfilling the IP requirement, including consortia and new or small investors (providing they themselves were members of such consortia capable of financing the whole project). This stage of the IP procurement was also used to ensure that all applicants being invited to negotiate and submit bids for the project satisfied certain minimum criteria, these being set by Thames Water after taking into account the requirements of the Government and Ofwat. The approach worked, with fifteen applicants registering to receive the PQQ letter, of which fourteen then responded prior to the deadline. Two rounds of bidding then followed, after which a 'revise and confirm' (R&C) bidding round took place.

For the first round, the strategic objective was to facilitate the bidders forming consortia capable of fulfilling the IP requirement, and to reveal their initial plans for financing the project and their initial thoughts on pricing. We believed it was also an opportunity to ensure the bidders were acquainted with the key features of a complex project before they were presented with the large volume of information disclosed in the data room in Round 2.[4] It provided the bidders with critical information that would facilitate their assessment of their ability to achieve an investment grade financing structure. Round 1 also provided an opportunity to down-select bidders if required, though the probability of down-selection being required was always considered to be low.

This bidding round was very successful, with all of the objectives being achieved in full. Thus, through this process, the fourteen pre-qualified Round 1 bidders coalesced into four potentially viable consortia, which engaged in the process and were ultimately expected to submit bids at Round 1. However, two bidding consortia chose to withdraw from the IP procurement shortly before submitting their Round 1 bids, leaving two bidders – these being the bidder that became Tideway and a consortia featuring Infracapital, USS, Borealis and Innisfree.

The main strategic objective of Round 2 was to elicit two ultimate bids capable of being accepted by Thames Water, the Government and Ofwat. Ideally, these would be of such good quality that no further negotiations would be required (to avoid delays) and on which the bidder had already undertaken full due diligence. Other objectives here also included the negotiation of some of the project documents in a competitive environment and the refinement and fleshing out of financing plans.

The key objective was fully achieved with two high-quality bids being received. However, as the Round 2 evaluation scores were virtually the same, it led to the decision to move to the final R&C bidding round (the decision to hold this took into account any potential time delays, the cost implications for Thames Water customers due to the delay and the views of the Government and Ofwat). This extra round did ultimately prove beneficial, however, saving Thames Water customers £142 million after the two bidders revised their costings.

After the R&C round, Bazalgette Tunnel Ltd (Tideway) emerged as the victor. Preferred Bidder status was awarded on 13 July 2015, with the licence granted on 21 August that year and coming into effect on the 24th. The decision was ultimately based on the quality of the bidder's financing plan as considered by investors, the rating agencies (which provided provisional investment grade ratings during the tender process) and the three stakeholders – Thames Water, the Government and Ofwat. Ofwat took the view that of the two bids, that of Tideway was the most robust and that best able to bear the risks allocated to the IP. The Weighted Average Cost of Capital (WACC) bid by Bazalgette Tunnel Ltd was 2.497 per cent, significantly below the water industry average of 3.7–3.8 per cent.* The competitive IP procurement had really delivered excellent value for Thames Water customers: it demonstrated that the investors would be prepared to accept a lower rate of return on water industry assets.

As a consequence of the IP procurement and the construction works procurement, the estimated bill increase was amended from £70–80 (which it had been for many years) to £20–25 per retail customer per year by the mid 2020s – a significant achievement.

As a final development in the IP procurement process, Ofwat held a second consultation in August 2015, when the regulator consulted on whether Bazalgette Tunnel Ltd was fit and proper to hold this key project licence to build the Thames Tideway Tunnel. Once again, these consultations were successful from Tideway's perspective.

Alignment and coordination between the different workstreams being managed by Thames Water and its key stakeholders (DEFRA, Infrastructure UK, Ofwat and other advisers) was achieved through a Senior Reference Group (SRG), which met in a non-executive capacity under the chairmanship of Mike Gerrard regularly throughout the project's development period.

In terms of developing the unique delivery model for the project and procuring the Infrastructure Provider, it was truly a fantastic team achievement. The team – led by my colleagues Amar Qureshi and Will Lambe, and including UBS, Linklaters, KPMG and Sharpe Pritchard – all contributed significantly and were instrumental to the

---

* The WACC is the regulated return that Bazalgette Tunnel Ltd earns between licence award and 2030.

success of the project prior to the appointment of Bazalgette Tunnel Limited. To mark their achievements, £1 coins that represented the consideration for transfer of assets that were sold to Bazalgette Tunnel Limited by Thames Water were presented by Thames Water's Chief Financial Officer Stuart Siddall to Amar and Paul Leece of UBS. As Mike Gerrard says: 'Tideway is yet another example of UK leadership in infrastructure delivery: it is the world's first construction project delivered as a stand-alone regulated utility.'

Bazalgette Tunnel Ltd (Tideway)

Now is an appropriate time to set out some of the detail of Bazalgette Tunnel Ltd. Tideway's shareholders are a consortium of investors, comprising funds managed by Allianz, Amber Infrastructure Group, Dalmore Capital Limited and DIF. The group, having been appointed the IP for the project, is financing around £3.1 billion of the Thames Tideway Tunnel project, with £1.1 billion coming from Thames Water. Tideway is the first waste-water-only utility in the UK, with customers in the Thames Water region paying it for the service that it will provide when the tunnel is constructed. The appointment of Tideway as the IP was subject to formal designation and licensing by regulator Ofwat, with the chief executive of the new company being Andy Mitchell and Sir Neville Simms chairing the new organisation. Tideway now has over 140 employees,* all dedicated to building this engineering masterpiece on cost and on time, aiming to improve the quality of life for all Londoners, and in the process reconnect London with the River Thames.

It is important to note that CH2M Hill had been the programme manager for the project since March 2008. By 2015, they had a significant team working on the Thames Tideway Tunnel, who between them had a huge knowledge of the project. In order to continue their crucial role, their contract with Thames Water was novated (transferred) to Bazalgette Tunnel Ltd. This was a critical initiative to ensure the successful delivery of the project and also provided reassurance to the new investors.

As the IP, Tideway is fully independent of Thames Water and has its own licence from Ofwat. The company is very specifically responsible for delivering the Thames Tideway Tunnel, with the

---

* As of September 2017.

IP duties in that regard including managing the contractors who will construct the main works and system integrator works that, once built, will be operated by Thames Water. The Thames Tideway Tunnel Project Specific Notice[5] defines the scope of the project to be delivered by Tideway the IP. The Thames Tideway Tunnel Project Preparatory Work Notice[6] sets out the relatively small parts of the project that Thames Water may deliver.

As the IP, Tideway specifically has responsibility for relationships with the following:

- Ofwat (through the project licence).
- Government in the form of DEFRA and the Secretary of State.
- Thames Water Utilities Ltd.
- Insurers.
- NEC project managers for the three main works contracts and the system integrator contract.
- The main works contractors (as mentioned above).
- The system integrator contractor.

It was decided much earlier in the programme to construct the Thames Tideway Tunnel in three contracts (West, Central and East) with an overarching systems integrator contract. With regard to these sections, the task of Tideway is firstly to make sure that each of the contracts is delivered successfully, but also to make sure that all the component parts work together to provide the optimum solution in one fully integrated hydraulic system.

A final point here, having covered the story of the IP procurement and providing some background regarding Tideway itself, is to reflect that, following this journey, we all here believe that our story is a potential blueprint for other major infrastructure projects around the world. One lesson worth setting out here, however, which will prove to be illustrative in subsequent chapters, is neatly detailed by Mike Gerrard when he says:

> When jumping hurdles, projects must temper ambition with experience. Set the hurdles too low and the speed of approach may not be enough to carry the project forward to the next hurdle; set it too high and a crash to the ground can prove fatal for the project!

A useful mantra as the story of Tideway develops!

5

THE BIGGEST CHALLENGE: ARGUING OUR CASE

I am passionate about every aspect of the Thames Tideway Tunnel. If you were to put me on the spot and ask me what the most difficult task was in the lengthy process of deciding on the solution, design development, formulating the unique delivery model or the complex procurement, and then initiating the construction process, I would answer 'none of them'. This even takes into account the complexity of threading this 25km tunnel beneath one of the world's largest and busiest cities, and specifically underneath (for the most part) one of the world's best-known rivers.

No, my answer would be that the most difficult aspect of the entire project was the consultation and planning application process (both the largest of their type in the UK to date), the former over two phases between 2010 and 2012 together with interim engagement and subsequent targeted consultations, and the associated communications programme.

The consultations involved thousands of individual engagements where we had to fight (as it felt at the time) tooth and nail to get our arguments across regarding the need for the Thames Tideway Tunnel to stop London's iconic river being used as an open sewer. With my allies Richard Aylard and Nick Tennant, and other colleagues from Thames Water and later Tideway, this engagement was part of a wider stakeholder engagement plan that began as soon as the tunnel project team was formed, and continues to this day. These engagements have taken us into the heart of Whitehall with the most senior political figures in the land, have involved explaining

93

our plans to royalty, have featured heated arguments with A-list Hollywood actors, celebrity chefs and the best-known comedians, and have taken in everything in between. Yet the most passionate responses we faced in our consultation process were actually from the Londoners whose everyday lives were to be affected by the construction of the tunnel and the CSO interception works.

From a personal perspective, even taking into account the wide variety of experiences I have had in my long career in the water industry, the stakeholder outreach for the Thames Tideway Tunnel was a massive step up in terms of the sheer size of the activity, particularly with regard to the jeopardy involved if we had failed at any one of the various stages. I had, of course, worked on major programmes in the sector before, including on two extensions to the Thames Water Ring Main, now a major part of London's water supply infrastructure. This ring main is around 80km in length, comprised mostly of 2.54m diameter concrete tunnels, and is used to transfer potable water from the water treatment works (WTW, note, not WWTW) in the River Thames and River Lee catchment areas for onward distribution to London. The core Ring Main (costing around £250 million) was built between 1988 and 1993, with the two extensions I worked on being added later. By way of comparison, the Thames Tideway Tunnel is currently estimated to cost £4.2 billion* (see Chapters 3 and 4). High stakes indeed, then.

Core Concerns

Over and above the other components of the London Tideway Improvements, the Thames Tideway Tunnel has always been controversial. This has largely been in the context of cost, whether the tunnel is in fact needed at all, whether it is the right solution in terms of sustainability and being environmentally friendly, and the physical disruption involved to Londoners during its construction. Despite this, we have always believed it to be the correct solution.

Since it was initially proposed, questions have frequently been raised about the cost of the scheme, given the current estimate of £4.2 billion and the fact that the project is being wholly funded by Thames Water customers. Some customers have been angered in this regard, especially as the scheme is a far more difficult one

---

* In 2014 prices.

to sell than its Bazalgette equivalent in the nineteenth century. This is because, at the time of the Great Stink, the dire condition of the River Thames and its tributaries was evident for all to see, and indeed smell. Today, however, while the environmental danger is there, it is less evident and it is its growing nature that has driven the need to take immediate measures to avoid another 'Great Stink'. Further, many of Thames Water's customers were/ are totally unaware of any issues with regard to the pollution of the River Thames and its tributaries, and indeed of any construction with regard to the Thames Tideway Tunnel taking place. Many Thames Water customers outside of London (see Chapter 1 for a geographical breakdown) question why they are having to pay for something they believe they will have no direct benefit from. The UK water industry model is that when infrastructure is required in a particular water company area, all customers of that company share the cost. Thames Water London customers have been paying for infrastructure constructed outside the capital for many years. So it is only fair that the cost of the Thames Tideway Tunnel should be shared by all Thames Water customers, whether they live within or outside London.

Meanwhile, in terms of disruption, those living alongside the proposed construction sites were, and indeed still are, understandably concerned about the noise and other types of disruption (such as road traffic and the possible loss of public space resulting from the construction of the project). One of our initial objectives here was to establish a Code of Construction Practice (CoCP), which would outline site-specific and project-wide measures to mitigate such concerns as much as possible. The CoCP implemented the best practice, which is now a feature across all of our sites, and covers: noise; transport (both by road and on the river); vibration; air quality; water resources; land quality; waste management and resource use; ecology and conservation; and the historic environment.

In terms of those opposed to the Thames Tideway Tunnel itself (and whether it is the most sustainable and environmentally friendly solution), sceptics have long maintained that a SuDS or other similar green infrastructure system would be a better and more sustainable way of dealing with the problem of London's sewerage system. Such a solution would, they argue, remove any need for the Thames Tideway Tunnel, given the alternative use of paved, permeable surfaces to replace the current impermeable paved surfaces across London, together with the use of green roofs, swales and water butts to promote the infiltration of rainwater into the ecosystem. This, they contend, would separate rainwater and waste water and hence solve the overloading of the combined sewer system. Such opponents of the Thames Tideway Tunnel solution not only believe these measures would provide the degree of control required, they also argue that they would be cheaper. Additionally, and set out here to provide balance, they say this green infrastructure would also: increase resilience across London to drought and flood events; reduce air pollution, always an ongoing issue in London; combat climate change; improve the enjoyment of Londoners of their environment, aesthetically and in terms of the health benefits of green spaces and the natural environment; reduce the urban heat island effect in the city and thus reduce carbon emissions through lessening the cooling load; improve urban biodiversity in the city.

The above points and arguments were all passionately made by our opponents throughout the consultation process. Clearly, we at Tideway believe that the Thames Tideway Tunnel solution still remains the best solution, both in terms of cost-effectiveness and also with regard to sustainability. In hindsight, at the time our arguments to promote the tunnel were somewhat polarised. We should have been clearer that SuDS and the Thames Tideway Tunnel are not mutually exclusive.

Consultation Overview

The consultation process, over two phases and a number of interim and subsequent targeted consultations of smaller size, was the vital component in the planning process for the Thames Tideway Tunnel. Of immense scale, it was our way of taking into consideration the views of all Londoners affected directly and indirectly by the construction process, and to a lesser extent the project's later operation. In the initial stages a letter was written to over 300,000 Londoners who either lived near a potential construction site or near the potential route of the main tunnel and connection tunnels. I signed these letters – a decision I later regretted, because of the sheer number of people who then had my contact details.

The overall process led to some very frank exchanges of views but, crucially, it allowed us to gauge an honest appreciation of people's concerns and then to take them into account when finalising the engineering solution for the tunnel and CSO interception works now being built. For example, we consulted on the actual physical route options of the main tunnel and connection tunnels at both the first phase and second phase of consultation, with three main tunnel route options then being under consideration. In fact, the final route was not actually locked into place until all of the views expressed in both consultation phases had been considered in detail. Further, all of the results of the consultation process were fed directly into the planning application that we were required to complete in order to obtain the DCO before any construction could begin.

For many of the public meetings, often in front of hostile audiences, the key team members who led the presentation of our case were Richard Aylard and myself, and I turn to Richard to set out some initial thoughts:

When we first started the consultation process in 2010, Phil and I set out two principles very early on. The first was that we would do all of the consultation ourselves as Thames Water and not hire in external consultants to do it for us, for example from the world of public relations. In that context he and I were the public face of the consultation, working alongside experts such as those designing the tunnels (and CSO interception works). In our view, if that meant that the public were listening and talking to engineers who were being a little technical, then at least they were talking to our team, the experts who knew what they were talking about. Secondly, we decided that we would not turn down a request for a meeting. We said to each other that we would sit down with everyone who requested a meeting, even if it meant visiting their own house, and in that sense the consultation was very hands-on. We knew what that meant in terms of the time commitment required, but took the view that it was the right thing to do. It meant some difficult and emotional conversations with people, but we didn't hide behind anything, we were as responsive as we possibly could be.

I can personally add two things that Richard and I, and our fellow team members, changed in the consulting process due to early learning experiences. Firstly, after our particularly challenging first public meeting regarding the King Edward Memorial Park Foreshore (KEMPF) site, we established our own rules of engagement with regard to such meetings. Thus, from that point we always fielded two project team representatives, insisted on an independent chair, and also on a venue agreed in advance with us. Secondly, as the vast majority of our meetings were in the evenings after long days in the office, we found that at the initial big events we were flagging by 9 p.m. due to a lack of energy. Thus, a good meal beforehand became the order of the day, where additionally the team could plan the evening ahead. These were great examples of our learning on the job as the consultation progressed, albeit driven by self-preservation.

Throughout the consultation process, 9,428 people/organisations responded to our outreach. This included twenty petitions (with over 40,000 signatures) that were generated by those arguing against aspects (some all, some just part) of the tunnel as a solution to London's current and future waste water problems. Of these, five were in relation to Phase One, four to the subsequent interim engagement, then nine with regard to the Phase Two consultation, and finally two petitions in relation to the post-Phase Two targeted consultation. The consultation process also included 135 advertised exhibitions, which attracted over 7,000 attendees, and we at Thames Water also received 2,600 customer calls with regard to the consultation. Further, we placed over sixty newspaper adverts to ensure Londoners were kept fully up-to-date with the consultation process, and participated in over 300 public meetings of all kinds.

Institution of Civil Engineers Open House Boat Tour, September 2014. The script included an explanation of the project to 300 members of the public.

Phase One Consultation

At Phase One Consultation we presented our shortlisted tunnel routes and sites and we identified which ones we preferred. We undertook a launch event with the key stakeholders that included the relevant local authorities. This took place in the lecture theatre at the Royal Geographical Society, Kensington. This event was noteworthy for a number of reasons. Firstly, it was a realisation for many of those who attended of the scale of the project, and the works proposed in their own local authority. Secondly, a number of those present (particularly Sir Merrick Cockell, Leader of the Royal Borough of Kensington and Chelsea) were very concerned that the power to approve the works in their area would be taken by others, namely the Planning Inspectorate.

The consultation was carried out between 13 September 2010 and 14 January 2011, lasting a total of eighteen weeks and involving the issuing of over 162,000 letters to Londoners who might be potentially affected by the construction and operation of the Thames Tideway Tunnel. The consultation resulted in a total of 2,864 respondents (408 after the end of the consultation period, a pattern that repeated through the whole consultation period, though these late submissions were still considered), with the five petitions containing 104, 401, 683, 5,013 and 5,274 signatories respectively.

For the Phase One Consultation a 170-page report was published, accompanied by a fifty-page summary and an almost 800-page supplementary report.

Interim Engagement

A combination of feedback from Phase One consultation and some sites no longer being available resulted in changes to some of the preferred sites. We wanted to give people the opportunity to comment in that regard before the Phase Two consultation. Therefore, we held an interim engagement on the sites that had changed from those presented in the Phase One consultation. This interim engagement was held between 11 March 2011 and 16 August 2011, lasting twenty-two weeks, involving informational mail drops to residences and locations within 250m of the construction sites. The most significant reasons for this interim engagement were to inform the public that we were going to consider Carnwath Road Riverside in the London Borough of Hammersmith and Fulham as an alternative to Barn Elms, and consider Chambers Wharf in the London Borough of Southwark as an alternative to King's Stairs

Gardens. This engagement resulted in a total of 311 respondents (thirty-one after the deadline, though again these were analysed and taken into account) and the four petitions detailed above, these latter containing 64, 120, 168 and 4,766 signatories respectively.

For this interim engagement consultation a 138-page report was published, featuring an additional thirty pages of addenda.

Phase Two Consultation

At the Phase Two consultation we again presented our preferred sites and tunnel routes in case there were new people to the area who had not previously had the opportunity to comment on them. We also provided more detail about the preferred sites and what we had done to mitigate impacts.

We had a launch event for the Phase Two consultation process at the Methodist Central Hall in Westminster. The event was very well attended, including politicians who were very keen to know our final decision about the main tunnel drive site locations, such as Greg Hands, MP for Chelsea and Fulham. There was a great deal of tension present in the room as I unveiled our site map, which

With Greg Hands MP, Stephen Greenhalgh (Leader of Hammersmith and Fulham Borough Council) and Kit Malthouse (Member of London Assembly), at a public meeting in relation to the Carnwath Road Riverside site in April 2011.

showed the twenty-four preferred sites selected to build the project. It is particularly noteworthy that we had taken on board the 9,400 responses we had received for the previous consultation and interim engagement, and where possible had modified our plans. Prior to the meeting starting, an opposition group had placed a leaflet on all of the seats in the vast hall, highlighting their views as to why the project should not proceed.

The consultation was carried out between 4 November 2011 and 10 February 2012, lasting fourteen weeks and involving the issuing of 129,000 letters. This phase of consultation resulted in a total of 6,010 official respondents and the nine petitions mentioned earlier, with a total of 6,553 responses of every kind being received overall; 333 were received after the close of the consultation period. Meanwhile, the petitions contained 233, 256, 291, 592, 1,115, 1,388, 4,766, 10,528 and 7,602 signatories respectively.

For this Phase Two consultation a report was published featuring a summary report of 74 pages, a main report of 558 pages and an extensive supplementary report of over 1,700 pages.

Targeted Consultation
A variety of bespoke post-consultations took place to give people the opportunity to comment upon aspects that had changed since those presented at the Phase Two consultation.

One targeted the following four sites: Barn Elms (access road), Putney Embankment Foreshore (site moved away from the bridge), Albert Embankment Foreshore (access road) and Victoria Embankment Foreshore (design). This took place between 6 June 2012 and 4 July 2012, lasting twenty-nine days and involving the issuing of 15,000 letters. This activity received a total of 135 respondents (12 after the close of the consultation period), together with the two petitions mentioned above. The latter contained 54 and 65 signatories respectively. Following completion of this targeted consultation a summary report was published, containing over 335 pages.

A consultation considered changes to site boundaries regarding access roads at the Albert Embankment Foreshore and the Falconbrook Pumping Station sites. This ran from 2 August 2012 to 5 October 2012, lasting sixty-five days and involving the issuing of seventeen letters.

A consultation considered changes to site boundaries regarding highway mitigation measures at King George's Park and Cremorne

Wharf Depot. This ran from 26 October 2012 to 26 November 2012, lasting twenty-eight days and involving the issuing of more than forty letters.

A sensitive equipment consultation ran from 29 October 2012 to 26 November 2012, a total of twenty-eight days. This involved the issuing of 450 letters.*

A Section 48 (of the 2008 Planning Act) Publicity consultation, part of the wider planning application process requiring engagement with, and a response from, local Government, ran from 16 July 2012 to 5 October 2012, lasting twelve weeks, and was publicised via newspaper advertisers. This process received a total of eighty-three respondents who provided feedback, of which three were received after the close of the publicity period.

A consultation considering changes to site boundaries regarding boat moorings and the replacement pier at the Victoria Embankment Foreshore and the Blackfriars Bridge Foreshore sites ran from 15 July 2013 through to 12 August 2013, lasting twenty-eight days and involving the issuing of over fifty letters.

Finally, a further targeted engagement consultation regarding minor changes to our application for development consent at the Victoria Embankment Foreshore and the Blackfriars Bridge Foreshore sites ran from 4 October 2013 through to 12 November 2013, lasting forty days (including weekends) and involving the issuing of over 740 letters.

From the number of letters mentioned, you can see that another huge undertaking with a project this size is what is known as 'land referencing'. This is determining who owns or has an interest in the land that the project will either use temporarily or acquire permanently. It was very important that letters and notices were sent to the right people.

The exercise included land affected by potential compulsory purchase and the alignment of the tunnel/shaft sites, as well as identification of those parties that may have had a relevant claim for compensation as a result of the scheme. Land referencing was undertaken through a process of diligent inquiry, undertaken by a specialist land referencing team, including preparation of a bespoke and tailored database for managing this scale of data, purchase and import of HM Land Registry information, desktop research,

---

*   'Sensitive equipment' is apparatus that could be affected by vibrations from tunnelling.

site visits and requests for land ownership information using pre-populated Land Interest Questionnaires. We conducted up to three site visits to each of 15,000 properties and recruited, trained and utilised a team of twenty-five throughout the ten-month period. As a result, a Book of Reference was produced and submitted to the Planning Inspectorate. It contained 4,338 pages and listed 56,000 land interests.

On requesting land ownership information from recipients across the scheme, and being wary of the potential impacts of vibration, we asked an additional bespoke question on the 'Tideway Land Interest Questionnaire' querying if any vibration-sensitive equipment was located on the land in question. Amongst various relevant responses, we additionally received responses noting the recipients' 'trees', 'cats' and 'me'.

Our site referencers gained a full appreciation of developments and ownership across a snapshot of central London, gaining an insight into architecture, cultures and society across the capital, experiencing first-hand the fast turnover of properties in an ever-changing environment. This work was undertaken by a very capable and diligent team at Mouchel.

Consultation in Action

Outcomes of the Consultation Process

In our initial activities of Phase One consultation, Thames Water reached out to all of the relevant authorities, organisations and individuals who would potentially be affected by the construction process of the Thames Tideway Tunnel, the aim being to obtain feedback on the preferred tunnel route options and the preferred locations of all of the sites.

In the case of the former, initially we had three potential tunnel routes being considered:

The River Thames Route, whose alignment broadly followed the river itself from west London through to the Beckton WWTW, cutting across the Greenwich Peninsula. The Greenwich Peninsula shortcut would reduce the length of the tunnel at a location where there were no CSOs that needed to be intercepted along the line of the river. This was the route referred to in DEFRA Minister Ian Pearson's letter of March 2007.[1]

The Rotherhithe Route, whose alignment was similar to the above River Thames Route but which would have cut across the Rotherhithe Peninsula in addition to the Greenwich Peninsula, thus reducing the length of the main tunnel by some 1.8km. However, this route would require longer connection tunnels from some of the key CSOs (this will be discussed in further detail in Chapter 6).

The Abbey Mills Route, different from the River Thames and Rotherhithe routes because it would connect the Thames Tideway Tunnel to the head of the Lee Tunnel at the Abbey Mills Pumping Station. This route followed the same route as the first two options until it deviated from the River Thames at KEMPF towards Abbey Mills Pumping Station. The main tunnel length of this route was around 9km (5.6 miles) less than the River Thames Route, saving some £700 million.

In terms of site selection, a longlist of 1,150 potential sites was developed by a desktop land surveying process looking for two types of sites: land either side of the thirty-four most polluting CSOs identified by the TTSS for CSO interception sites, and sites either side of the River Thames for locations to build the main tunnel. These were then further examined against a variety of set criteria covering five disciplines, including planning issues, engineering difficulty, environmental impact, and local property and community considerations. The longlist was filtered down to a draft shortlist, then a final shortlist to arrive at a preferred list. At each filtering stage the level of detail considered by each of the five disciplines increased. This multi-disciplinary and multi-stage approach involved a wide-ranging team and particular thanks goes to Patricia Stevenson, whose commitment to managing the huge site selection process was central to its success.

Three site selection methodologies were drafted before being finally agreed with the fourteen relevant London local authorities. This proved to be an incredibly useful approach, because although the local authorities could disagree with the selection of a site, it was imperative that they could not disagree with the methodology used to select it.

The three tunnel routes and the final shortlist of sites showing which ones we preferred then formed the basis for the Phase One consultation process that, as set out above, took place between September 2010 and January 2011. Hydraulic modelling of the sewerage network and a review of how the CSO discharges could be controlled determined that not all thirty-four CSOs needed to be intercepted. There were twenty-three Phase One consultation preferred sites: three were main tunnel only sites, three were combined main tunnel and CSO interception sites, and the remaining seventeen sites were CSO interception only sites.

In response to the comments received in this phase, radical changes were made. Firstly, the main tunnel drive site in the western section was changed from Barn Elms to Carnwath Road Riverside, and the eastern drive strategy was changed to drive from Chambers Wharf in the London Borough of Southwark to Abbey Mills Pumping Station. Secondly, hydraulic modelling determined that a further three CSOs could be controlled without intercepting them but by modifying the local network. This was very good news as some particularly controversial sites (for example, that planned for Druid Street) were removed. Thirdly, where the sites were to remain in the same location, additional major improvements were made. As one can gather, a massive effort was made to listen to the issues and concerns raised, and where possible to amend our plans.

Some of the elements of this revised set of options were then presented to stakeholders in the Interim Engagement, which took place between March 2011 and August 2011. As part of this process the residents who lived within 250m of the eleven specific new preferred sites were sent the informational mail drop, set out above, which explained that the sites were being considered as alternatives to the original preferred sites. The mail drop invited residents to attend drop-in sessions where they could pose questions and, having listened to the response, gain a better understanding of the overall project and how it might affect them. As part of this process, in total ten two-day sessions and one larger community liaison meeting were held. These sessions and the meeting were attended by over 800 people, and in all 168 comment cards and 147 pieces of site-specific correspondence were received and then considered.

Based on this Phase One consultation and Interim Engagement process, Thames Water (in consultation with the Government, EA and Ofwat) decided that, for the overall project to be as cost-effective as possible while causing the least disruption and yet still meeting the EU's UWWTD requirements, the following components were to make up the preferred scheme for the Thames Tideway Tunnel:

- The shorter Abbey Mills Route tunnel comprising a 25km main tunnel of 7.2m diameter (except between Acton Storm Tanks and Carnwath Road Riverside where the diameter is 6.5m).
- Two long connection tunnels and nine short connection tunnels.
- The twenty-four preferred sites comprising: four main tunnel sites; one main tunnel and CSO interception site; sixteen CSO interception sites; two sewerage modification sites; one modification site at Becton WWTW.
- The thirty-four CSOs controlled as follows: eighteen intercepted (two CSO sites each intercept two CSOs and one CSO requires two CSO sites); two local sewerage modifications; fourteen indirectly controlled.

This post-Phase One consultation/Interim Engagement preferred route site list was then presented for the second round of public consultation, with a broader set of questions being asked of stakeholders.

In the first instance, this was in the form of the Phase Two consultation, which was carried out between November 2011 and February 2012, engaging local authorities, landowners, local businesses and communities. These were consulted on the following: the overall need for the Thames Tideway Tunnel project at all, and whether it was the most appropriate solution to the requirement regarding London's future waste water provision; their views on the preferred Abbey Mills Route for the tunnel (including its detailed alignment); their views on the preferred sites and permanent works as selected following the Phase One consultation and the Interim Engagement; the effects that the Thames Tideway Tunnel project would have on the local environment, as outlined in the Preliminary Environmental Information Report developed following the Phase One consultation and Interim Engagement.

Following this consultation process, the Abbey Mills Route was selected and twenty-four sites were proposed to go into the planning application. However, several sites were identified as requiring further, targeted consultation (each individual one set out in the preceding overview section) seeking comment upon changes made since the Phase Two consultation. This then resulted in a further refinement process before the overall scheme was signed off by Thames Water, with the final exact route and site list being set out in Chapter 6, which covers the ultimate engineering solution.

The site selection process, together with the feedback from the consultation exercises, drove our land acquisition strategy. It was imperative that we secured the parcels of land required to build the project as soon as we could. A key part of this process following the start of the Phase Two consultation was to ensure that DCLG 'safeguarded' (by way of a Safeguarding Direction) the land that was required. We then legally became a statutory consultee as part of the planning process to develop such land. It was our strategy, wherever we could, to buy the land required through negotiation, no land to date has been compulsory purchased. We were very successful in acquiring the required land and hence providing security for the now proposed scheme.

The key land parcels were those needed for the main drive sites at Carnwath Road Riverside, Kirtling Street and Chambers Wharf, all of which are interesting in their own right.

Firstly, Chambers Wharf became available after the launch of the Phase One consultation when St Martins (owned by the Kuwait Government) decided to sell all their land holdings in the UK, having started the residential development on the site, for which they had planning permission. The purchase of this land was key to us moving from the greenfield site at King's Stairs Gardens to the brownfield site at Chambers Wharf. Prior to us going ahead with this £78m acquisition I had a meeting with Annie Sheppard, CEO for the London Borough of Southwark, to obtain her support for us purchasing the land from St Martins. She agreed to support this course of action.

Secondly, the site area needed at Kirtling Street required the purchase of seven parcels of land, mainly from developers, and one of the most protracted acquisitions, namely the purchase of the Cemex site, which led to the relocation of the company's concrete batching plant. Cemex were reluctant to move the plant, which had river access, and worked hard to maximise their compensation payment.

Thirdly, the land required for the Carnwath Road Riverside site was made up of three parcels of land: Whiffin Wharf, Hurlingham Wharf and the Carnwath Road Industrial Estate. The most significant challenge was to purchase the freehold of the industrial estate from the London Borough of Hammersmith and Fulham. Firstly, the site was a major component of the land we required; and secondly, due to our strained relationship with the London Borough of Hammersmith and Fulham, I concluded the deal myself with Cllr Stephen Cowan, Leader of the Council.

King's Stairs Gardens Public Meeting at The Bosco Centre, Bermondsey.

Given the length of the process (September 2010 to November 2013) and the detail involved, together with the passions shown by all of those with a view to our storage and transfer tunnel solution for London's future waste water requirements, we had a lot of colourful experiences. In the end, as with many things in life, a great deal came down to the personalities involved.

Barn Elms and Putney Embankment Foreshore

The Barn Elms initial main tunnel drive site and ultimate CSO interception site, together with the nearby Putney Embankment Foreshore CSO interception site, have a strong historical heritage that is keenly felt by the local communities – for example, Elizabeth I's spymaster, Sir Francis Walsingham, used to reside at Barn Elms Manor. Meanwhile, adjacent to the site at Putney is St Mary's Church, the location of the human rights-related Putney Debates during the English Civil War. Against such a cultural backdrop, many local residents questioned the actual need to connect the associated CSOs to the Thames Tideway Tunnel. 'Stop the Shaft' were the principal catalyst behind local opposition at Barn Elms. They were specifically against a main tunnel drive site being located in a major west London area of Metropolitan Open Land and well-used sports fields.

Specific engagements, of which there were many, included our meetings with the Putney Working Group, where our plans at these locations were discussed with the local MP Justine Greening (who chaired a number of the meetings) and other local stakeholders. The meetings were held regularly from 2011 and were an exemplar of how a major project should interface with a local community. Justine Greening did an excellent job in inviting local community and business leaders to participate and we came up with a solution that really met the needs of local people. It was the only example where members of the group visited our offices at Paddington and sat down with the designers, working together to come up with solutions that were acceptable to all. Another was our public exhibition regarding the Barn Elms location on 14 June 2012 at the Wildfowl and Wetlands Trust Wetland Centre at Queen Elizabeth's Walk in SW13, where we attracted many local residents.

One of the most surreal experiences in the whole engagement process was the meeting at St Mary's Church Hall in Putney, with Graham Stevens of Bluegreen UK and a very tolerant Justine Greening, to discuss blue-green alternatives to the project and how nanotechnology would soon make our solution redundant. We concluded that said nanotechnology was still being developed, and is yet unproven.

Other fond memories include a certain Putney resident who made a video to demonstrate that the Carnwath Road Riverside protestors were exaggerating their traffic issues on Carnwath

Talking at the Putney Working Group meeting with Justine Greening.

Road, which she brought in to show me. At a public meeting in Putney, the same lady shouted out that Richard Aylard and I were the 'two guys with the flash watches'. Also Sian Baxter, a Barnes lighting specialist, who showed great courage by standing up at the Carnwath Road Riverside public meeting to put the case as to why the western drive shaft shouldn't be in Barn Elms playing fields.

We also attended a wide variety of other public meetings. These included a very large gathering with hundreds attending and chaired by Zac Goldsmith MP, which proved to be particularly hostile; and one in the Church Hall in Putney, where the information technology support failed and Richard Aylard and I had to perform live without the slide show we had planned. It involved lots of hand and arm movements.

Carnwath Road Riverside

The Carnwath Road Riverside main tunnel drive site is in Fulham. This is again a site with a strong local cultural resonance; for example, being the location of the late nineteenth- and early twentieth-century Metropolitan Asylum Board (MAB) ambulance service that, amongst many other activities, used a river facility here to transport infectious patients to isolation hospitals near Deptford. The MAB can also claim to have provided the first state-run hospitals in Britain, thus being the forerunner of the modern NHS and so having a very strong resonance with current public affairs.

Here we faced a number of organised opponents, for example the Residents Against the Thames Sewer (RATS), which evolved (through the addition of the Fulham Society) into Stop Them Shafting Fulham.[2] Local residents' organisations were also opposed to our activity here, for example that representing Peterborough Road. Broadly, those such as RATS and elsewhere were opposed to our activities at this site on the grounds that there would be three years of construction, twenty-four hours per day, which would cause noise, dust and light nuisance to nearby residents. There was also concern over traffic congestion caused by vehicle movements. Finally, there was concern over the effect on the five nearby schools. The project team believe that all these effects could be mitigated to a level that minimised nuisance to an acceptable level.

Some of our memorable activities included the drop-in session we held at the Hurlingham and Chelsea School on Peterborough Road on 7 April 2011 for concerned local residents and workers.

Here we hosted ninety-two individuals in total between 4 p.m. and 8 p.m., with over 580 people registering an interest.

There were also a number of very challenging and emotionally charged public meetings at Peterborough School, Fulham, in relation to the Carnwath Road Riverside site. At one particular meeting in 2011 attended by 350 people we were on stage with Stephen Greenhalgh, Leader of the London Borough of Hammersmith and Fulham, local MP Greg Hands and Kit Malthouse, Member of the London Assembly. The most notable points of the meeting were that Stephen Greenhalgh had the whole audience chanting in opposition of the project. I was also accused by Greg Hands of not telling the truth about the content of the Thames Water Bill leaflet. Although he was adamant he was correct, he was not.

A third very interesting experience, also in 2011, was my visit to BBC's Broadcasting House headquarters in Portland Place to participate in Eddie Nestor's BBC London drive-time show. Members of the public were invited to ring in to raise their concerns, and I was given the opportunity to respond. The show host was friendly towards me and at one point accused a protestor of being a 'nimby'!*

Chambers Wharf

The Chambers Wharf main tunnel drive site in Southwark. Historically, this site was originally riverside wetlands on the northern margins of Bermondsey eyot, west of the mouth of the River Neckinger, a Thames tributary. The modern Bermondsey Wall marks the line of the medieval river wall built by Bermondsey Abbey to protect local land from flooding, while later the post-dissolution estate of the latter was purchased and built on by Sir Thomas Pope, founder of Trinity College Oxford and associate of Sir Thomas More. More recently, the site has featured intense mercantile activity.

Here we struggled to an extent with our engagement with the local community and politicians, latterly almost as a result of the success of our consultation process. This was because the initial location of the start of the eastern section of the main tunnel was a reception site at King's Stairs Gardens, but we moved this to Chambers Wharf (a brownfield site) after listening to responses from the local community. However, this then led to a long campaign against this second location by the local MP Simon Hughes (at the

---

* 'Not In My Back Yard' – the characterisation of opposition by residents to a development in their local area.

time Member of Parliament for Bermondsey and Old Southwark, who had also opposed the King's Stairs Gardens original location) and other protagonists who objected to the new location, given it was close to local residences. The well-organised opposition here included SaveYourRiverside (SYR, chaired by Rita Cruise O'Brien with Barney Holbeche as the technical lead) and Tempus Wharf Freehold Management (led by Derek Joseph). SYR have been particularly vociferous in their opposition to the site being used as a drive site for the eastern section of the main tunnel, saying at the time of the granting of the DCO in September 2014:

> SYR condemns the Government's disgraceful decision to override its own planning inspectors by authorising the use of Chambers Wharf as a drive site for the Thames Tideway Tunnel. The inspectors agreed with SYR and (the London Borough of Southwark) that CW as a drive site was not justified due to adverse local impacts on residents and that alternatives should be looked at.
>
> Ministers contradict themselves. They claim on the one hand that the residual impacts on the quality of life and health from noise 'are less than significant' due to mitigation measures, but at the same time concede that noise and disturbance impacts weigh against consenting the development and that 'the impacts at Chambers Wharf will be adverse and of long duration'. They then go on to give consent nonetheless. This is manifestly disingenuous, unjust and an insult to local residents who are being made sacrificial victims for the project when the inspectors, who actually heard the evidence, agreed that less impactful alternatives are available.
>
> It does the credibility of the major infrastructure planning regime no good in the eyes of the public for proposals that were shown to be flawed in preparation, and then shown to be half-baked when tested at the Examination, to nonetheless be given consent by ministers with minimal regard for the impact on Londoners unfortunate enough to have to suffer for what is perceived as the wider good. Ministers have broken the faith by which people impacted by major projects could at least expect to be treated fairly and sympathetically, and given the green light to a company notorious for its corporate arrogance to do its worst to Londoners. This controversial project has just become significantly more controversial, a decision which will come to haunt politicians, officials, and Thames Water and its owners and potential investors.

Richard Aylard, Sir Patrick Stewart and Val Shawcross, Member of London Assembly, at the start of a public meeting about the Chambers Wharf site.

Site visit with Simon Hughes MP, a representative of London Borough of Southwark and members of SaveYourRiverside, to outline the difficulties of driving the eastern section of tunnel from Abbey Mills Pumping Station to Chambers Wharf, in 2013.

There were many engagements with stakeholders over the Chambers Wharf main tunnel drive site, but a key one was on 25 June 2012 when we held a public meeting at City Hall in London. This event was attended by over ninety-five people, including SYR, with a Thames Water team of three facing particularly tough opposition. At another Chambers Wharf public meeting there was a significant amount of challenge regarding the potential noise from the proposed barging operation. Rather unfairly, the actor Sir Patrick Stewart accused me of being 'absolutely outrageous'.

Such levels of opposition continue to this day, there was a call to supporters to attend a public meeting on 21 April 2015 to express their concerns to Tideway CEO Andy Mitchell and the then potential contractors who would be actually carrying out the work.

King Edward Memorial Park Foreshore (KEMPF)

The King Edward Memorial Park Foreshore CSO interception site, I feel sure Richard Aylard and Nick Tennant would agree, was the location of some of our most bruising encounters in the whole consultation process.

Historically, this area has long been associated with trade activity, being adjacent, for example, to the mercantile settlements at Shadwell and Ratcliffe and later seeing the expansion into the area of the East and West India trading companies. It was here where we once again ran into the world of celebrity, with Richard, Nick and I able to highlight such globally recognised actors as Helen Mirren, celebrity chef Delia Smith and comedian Lee Hurst, who all became involved in the campaign against the use of this location in some way. The latter on one occasion used some colourful language to make it very clear he didn't want KEMPF used as a site. Opponents of the project often used the support of such celebrities to further their aims in relation to specific sites, particularly through negative media attention.

At the first public meeting, I spent over one and half hours being shouted and screamed at. I was given very little time to respond to the questions raised; all those present were very keen to impress upon me that they did not want a construction site in the park. Throughout my time on the project, this was the only occasion when I felt that my personal safety was at risk. Following this event, the project team employed plain-clothes security guards at all future meetings in relation to this site. We also ensured that we had cars with drivers parked outside the venues so that, if needed, we could make a quick getaway.

In relation to the Phase One consultation, there were threatening tweets regarding a proposed public meeting that caused particular concern with Thames Water's CEO. As a result of the threats made, we had security checks undertaken in relation to the individuals concerned. Following comments regarding this made at the public exhibition in Shadwell, the chairman of the very active SaveKEMP group then wrote to the *Evening Standard* and an article was published accusing Thames Water of 'Big Brother' tactics. Also, due to sensitivities at the same exhibition, there was a kerfuffle when a lady was refused admission with her gardening shears!

Here we had engagements with a wide variety of organisations, including SaveKEMP, with whom we had regular meetings as part of our engagement with the London Borough of Tower Hamlets. For a long period these meetings were held on a monthly basis. Meanwhile, we dealt with a variety of other types of stakeholder during this process too, for example the Free Trade Wharf Management Company (latterly represented by solicitors Bircham Dyson Bell) and the Glamis Residents Group. With regard to the latter, representative Desmond Ellerbeck submitted a relevant representation, which is instructive as to the wider issues in play here:

> I represent the residents of the Glamis estate which is within 50 metres of King Edward Memorial Park. The estate consists of over 360 dwellings and over 1000 people. Many residents have families who use the park on a regular basis, and are extremely against Thames Water disrupting their amenity space when there is a viable alternate site at Heckford Street a few minutes away. The park is the only large green space in the area, and used by a large proportion of residents in an extremely densely populated part of London. This will blight Shadwell for years while the works take place.

Other organisations with whom we had engagements over time regarding the KEMPF site included the Old Sun Wharf (OSW) Freehold and Management Limited, also represented by Bircham Dyson Bell, and the Trafalgar Court Residents Association. Located very near to the KEMPF, the latter said in their own representation:

Receiving a petition against using King Edward Memorial Park as an interception site from local schoolchildren in February 2012.

> [The] Trafalgar Court Residents Association represents the inter-
> ests of owners and tenants who live and work in a block of 68 flats
> a short distance south of the proposed works in King Edward VII
> Memorial Park ... The impact on our daily life is measured in years
> and, therefore, we vehemently oppose the plan.

An activity that did not happen in relation to any other site was the invitation to a number of schools in the vicinity of King Edward Memorial Park to receive petitions from the schoolchildren. It was insightful to listen to the pupils' concerns and slightly worrying that they had only been informed of one side of the argument, about why the site was chosen to intercept the storm relief sewer through the park. I did my utmost to be empathetic.

It was with regard to the KEMP Foreshore CSO interception site that we carried out a disproportionate amount of our work designed to limit opposition to our work across the whole Thames Tideway Tunnel project. We held a consultation exhibition at the

John Scurr Community Centre on Bekesbourne Street, which took place on 18 October 2010 as part of the Phase One consultation. Twenty-five people attended, with the agenda focusing specifically on the KEMP Foreshore CSO interception site.

We then had a meeting with Andrew Boff, Conservative Member of the London Assembly, in City Hall on 15 March 2011. Here he raised a number of issues that are instructive to record, as they provide insight into many of the political issues raised during our consultation from such stakeholders. These included concerns regarding the access to green space, how the site was selected, any permanent structures on the site, and whether there even needed to be a site at KEMPF at all.

Finally the King Edward Memorial Park Foreshore public meeting was held on 12 January 2012 at St Paul's Church, Shadwell. With Jim Fitzpatrick (Member of Parliament for Poplar and Limehouse) in the chair, we fielded a strong home team here of six led by myself and Nick Tennant, and facing a hostile barrage of questions from those such as SaveKEMP. As always, however, we strove to answer any inquiries honestly, and by this time had become adept at spotting those who were keen to participate but wary of speaking publicly. One of our learnings as our programme progressed, we therefore

King Edward Memorial Park Foreshore Public Meeting at St Paul's Church, Shadwell, chaired by Jim Fitzpatrick MP.

always attempted to speak to such individuals at the end of any given meeting, when they felt less peer pressure.

Jim Fitzpatrick paid tribute to me and my team in the House of Commons on 29 February 2012 for work we did in relation to engaging with the community, listening to concerns and changing our plans for the proposed site in King Edward Memorial Park Foreshore.

Planning Application

While engagement continued with stakeholders into 2013 and beyond after the major component parts of the consultation process had been completed, by October 2012 the final deadline for Thames Tideway Tunnel's Section 48 process had closed, and with it the last real opportunity for stakeholders (including the public) to have their say on the updated proposals for the tunnel route and sites. The extensive application for development consent (by which the DCO would ultimately be granted) to construct and operate the Thames Tideway Tunnel was then delivered to the Planning Inspectorate (who would consider it on behalf of the Government) on 28 February 2013. This was a huge logistics exercise in its own right, given that our enormous document had to be delivered from our headquarters in London down the M4 motorway to Bristol.

The application consisted of 50,000 pages, of which the Environmental Statement made up 25,000 pages, and the hard copy required a 16m-long shelf to store it. To make its delivery even more problematic, the Planning Inspectorate required two hard copies of it. Each of these copies weighed around one tonne and had to be delivered in separate Ford Transit vans. One of my greatest, and most cathartic, memories of the whole Thames Tideway Tunnel story was riding shotgun for this delivery to the Planning Inspectorate offices (with Ian Fletcher and Rick Fornelli), where we were greeted by their chief executive Sir Michael Pitt, who expressed his surprise at the size of the application. The Planning Inspectorate then had twenty-eight days in which to decide whether our application was valid, and whether the huge variety of consultations we had undertaken were adequate to support the DCO application.

This happened quickly, and on 27 March 2013 it was confirmed that the application was indeed valid and that Thames Water's extensive consultation process for the project had been adequate with regard

Me, Martin Baggs and Jim Otta, at the event when Martin Baggs signed the DCO document prior to submission, in January 2013.

Delivering the two hard copies of the DCO documents to the Planning Inspectorate in Bristol, 28 February 2013.

With Rick Forneli, signing the DCO document boxes prior to submission, in January 2013.

to all requirements. At that point, all of the DCO application documents were then made available in their own specific section of the Planning Inspectorate's National Infrastructure website. Continuing our policy of openness with regard to our plans for the Thames Tideway Tunnel, Thames Water then also made the documents available for scrutiny at six public locations spread out along the proposed tunnel route, three on either side of the River Thames.

On 3 June 2013, it was announced that the Secretary of State at the DCLG, Eric Pickles, had appointed five inspectors as the examining authority to consider any matters arising. These individuals were Jan Bessell, Libby Gawith, Emrys Parry, Andrew Phillipson and David Prentis. As part of this process all interested parties were then invited to make written representations. Next, a preliminary meeting open to all of those who had registered an interest was held on 12 September 2013 at the Barbican Centre in the City of London. This was chaired by the Planning Inspectorate, who determined how the examination would be carried out, including any consideration of more detailed hearings on any site-specific matters as well as any other project-wide issues. We were pressed by the chair to state that we would be able to do everything required during the examination process, which was a very difficult question to answer. Our answer was obviously 'yes'.

Mike Gerrard, Richard Aylard, me and Jim Otta with a copy of the DCO document at the staff celebration of the DCO being submitted.

Calm Under Fire winner at the DCO Awards in April 2013. Award being presented by Mike Gerrard, Managing Director, Thames Tideway Tunnel.

All of the category winners at the DCO Awards in April 2013.

During the examination period there were forty-eight hearings. I attended forty-five, which were held at locations across London. There were three types of hearing: site-specific (relating to our different construction sites), topic-related (such as noise or transport) and open hearings (for geographical areas across the riparian local authorities where anyone could attend and ask questions). The noise hearings lasted for four days and the transport hearings for five days. My most intriguing hearing was in relation to Albert Embankment Foreshore, which was held behind locked doors at the Royal Courts of Justice. I led our negotiations with the Secret Intelligence Service (MI6). The only comment I can make is that, in order to undertake this role, I had to sign the Official Secrets Act!

The examination period, however, provided me with the greatest challenge of my career.

As the key witness for Thames Water, the promoter, I was called upon to answer questions about the development consent application and make statements and comments on behalf of the project team and Thames Water. Frequently I had to think on my feet and provide answers during the cross-examination. There was no time for me to check with others: it would have undermined the process and confidence in the promoter. It often felt a very lonely place. Some of the commitments I made have far-reaching implications for the project to this day, particularly in relation to the River Transport Strategy and noise.

In a number of instances we were given a strong steer from the inspectors that what we had proposed in the development consent application was not acceptable, and that if we did not commit to change our position we were at significant risk of not being granted a Development Consent Order. One such instance was our decision to change the River Transport Strategy to bring concrete tunnel segments into Chambers Wharf by river, and not by road, and to commit to 'no night time' barge movements at Chambers Wharf. On a number of occasions senior management meetings were convened at very short notice near the examination venues late in the evening to seek agreement for decisions where there would be significant implications for the project. There were long and some-times heated debates about the ramifications of changing the River Transport Strategy. The meetings proved to be a key test of my influencing skills: in many instances, the commitments and changes I made had multi-million-pound implications.

The site-specific hearings in relation to Chambers Wharf were the most challenging. The action group SaveYourRiverside were con-vinced that we had chosen the wrong eastern tunnel drive strategy. They believed we should drive the eastern section of tunnel from Abbey Mills Pumping Station to Chambers Wharf, rather than vice versa. The implication of this change was that there would have been far less construction work at Chambers Wharf. We were robust in the defence of our approach, which majored on the security of transporting materials to and from site via the main River Thames in Southwark, rather than using the very restricted River Lee from Abbey Mills Pumping Station. SaveYourRiverside had employed their own barrister and eminent noise expert. There were many hours spent debating our analysis of the predicted amount of noise that would emanate from the site during construction and how it would be controlled. There were concerns expressed about the effect the noise would have on local residents. I rapidly learned how difficult it is to quantify noise and about the different methodolo-gies for measuring it.

The hearings were exhaustive and involved a tremendous effort from many people in the project team. For each site-specific and topic-related hearing a 'control room' was set up at the venue with topic experts and links to our document management system so that we could rapidly access any information that was required.

Many of the hearings were on consecutive days and a number started at 10 a.m. and went on until the evening. It was hugely challenging to undertake a 'wash up' of what had happened during the day whilst having to prepare for the next day, which frequently related to a different site or topic. There were many nights burning the midnight oil. Throughout the process, Michael Humphries QC and James Good (Berwin Leighton Paisner solicitor) did an outstanding job in outlining and defending the project's case. John Rhodes, our planning expert, also played a critical role at the hearings and in answering the written questions raised. He was a tower of strength. In hindsight, managing the examination process was a monumental task.

Of the forty-eight hearings, five hearings were 'open'. These were immensely challenging to me as at the end of the hearing, on behalf of the promoter, I was given the opportunity to respond to any points raised. It was particularly important to not let any incorrect statements about the project stand unchallenged. I had a team of individuals rapidly putting together our responses. As one protestor put it, 'it relied on my encyclopaedic knowledge of the project and the water industry'. The points raised in the open hearings were incredibly wide-ranging and often covered areas outside the remit of the development consent application, such as other potential options like sustainable drainage systems.

Towards the end of the examination hearings it was evident that we were going to struggle to answer all the questions raised, to an appropriate standard, by the time the examination period expired. After great debate with the project team and Thames Water directors, it was agreed that we would write to the Secretaries of State for DEFRA and DCLG and ask for an extension of time. In the event, the Secretaries of State turned down the request and we worked exhaustively to provide the answers required by the deadline.

During the examination hearings we had to answer 630 written questions. Some answers to these questions were over 100 pages in length, and during the examination a further 75,000 pages of supporting documentation was submitted. This meant that in all, including the original development consent application, a total of 125,000 pages were submitted, which makes the Thames Tideway Tunnel project the largest planning application ever made in the UK.

Once this Inspectorate-led six-month-long examination activity had concluded, the examining authority was then able to make its positive final recommendation to the Government by confirming their approval for the DCO application, though saying that the case for Chambers Wharf as the eastern main tunnel drive site had not been fully made. This determination was then passed on to Government ministers to make their final decision, who then announced the granting of the DCO on 12 September 2014, in doing so overriding the Planning Inspectorate's concerns regarding Chamber Wharf. This was the happy end of a very long journey for all of us involved in the project, with the final DCO as granted being as clean as we could have hoped for. There was now a green light for work to begin.

Political, Media and Other Stakeholder Engagement

Although the project has always had broad cross-party support (see below), this was never taken for granted. We produced a stakeholder map for the project that was large enough to cover an entire office wall. We worked tirelessly to make sure key political stakeholders were given regular updates on the project, and paid close attention particularly to political newcomers at local level.

From the moment the TTSS recommended the single storage and transfer tunnel solution as their preferred response to London's current and future waste water requirements, we have been engaged in a significant outreach programme to all political, media and other interested decision-taking and opinion-forming stakeholders whose views have an impact on the success of the Thames Tideway Tunnel.

Broadly, in our engagement with all stakeholders there were two goals we strove to achieve with everything we did. Nick Tennant:

In the first instance authenticity. Dealing with often hostile and suspicious audiences was a real challenge, especially in the public meetings, with many assuming we would just tell people what they wanted to hear and then never return. This was of course not so, we always put our authenticity at the centre of all we did, trying to answer every question honestly. This meant that over the long term people began to trust us, even if begrudgingly, even when they still disagreed with us. And secondly being persistent, even today. Our communication with stakeholders of all types was and is a huge logistical exercise, and we decided early on that maintaining the momentum of our outreach activities was vital.

We also decided to break down all of our collateral documents used to support our communications outreach into easily accessible and understood bite-sized chunks, given the complicated messages and information we were trying to impart.

Political

Canvassing the support (or mitigating the opposition) of political decision takers and opinion formers ranging from national Government, Westminster in both its legislating and local Parliamentarian capacities, Whitehall (principally DEFRA and the DCLG), local Government and also Non-Governmental Organisations (NGOs), has been a central part of our daily lives for years now as we have campaigned for the Thames Tideway Tunnel project. Most significantly, it was of course Government that was the ultimate decision taker in granting our DCO, and then supporting the project on an ongoing basis, while all other political audiences have had their own role to play as our story has unfolded.

Unlike many other major infrastructure projects, the Thames Tideway Tunnel was unique in that it lacked a political champion, making the political component of the stakeholder outreach programme particularly important. As Mike Gerrard says: 'Tideway defied gravity by happening despite the absence of a strong political champion willing to speak out publicly, strongly and often in support of the project.'

A special mention, therefore, needs to be made of John Bourne, Deputy Director, Water Availability and Quality at DEFRA. With no political champion, it was critical that the project at least had a champion in Whitehall. John was that person, working on the project from 2009 to 2015. His positive attitude and determination, together with his pragmatic approach, were key to the project progressing and the delivery model becoming a reality. He was also a real gentleman and a pleasure to work with.

We have always maintained the closest possible contact with national Government, from the Prime Minister (four since the TTSS published its first report) downwards through the political chain of command. In fact, it is the transitions between Governments that have presented us with our most problematic situations politically, especially when the transition involved a change in the party of Government at a crucial time. As Richard Aylard says:

> The most difficult time was when the Government changed in 2010 from the Labour Government where Ian Pearson (Minister of State for Climate Change & Environment at DEFRA) had been a strong supporter and advocate, to David Cameron's Conservative/Liberal Democrat coalition. Obviously a new Government coming in wanted to have their own look at major capital expenditure projects such as the Thames Tideway Tunnel, with cancellation of the programme the worst case scenario. Fortunately we argued our case well and things progressed with the new Government being ultimately supportive.

To my recollection, this review of the Thames Tideway Tunnel by the Coalition Government was very thorough, with the newly responsible DEFRA Minister Richard Benyon MP examining matters from all angles, including engaging with some of the less supportive NGOs. As Richard Aylard says, however, we persevered and won the day, with the new minister becoming a strong and fair supporter.

It wasn't, of course, only such important political decision takers who were key to our political success, but the opinions of those around both them and, always, the Prime Minister. One of our key targets was Oliver Letwin MP, the Minister of State for Government Policy in the Cabinet Office in the Coalition Government. He had (and indeed has) a reputation for being a free thinker and for being the Prime Minister's troubleshooter, and word reached us early into the life of the new Government that he was taking a keen interest in the Thames Tideway Tunnel, apparently being sceptical of its need. We were having trouble engaging with him directly, and so Nick Tennant and I came up with a means of ensuring we had some time with him, with a view to arranging a longer meeting. Specifically, this was to use a Conservative Party Conference fringe event at the 2012 conference (we were and still are regular attendees at all major political party conferences: I've attended eighteen to date) where he was speaking as a means of engaging with him, and we struck lucky. Both Nick and I arrived at the given fringe event early, and as it happened so did the minister, alone and without his civil service minders. Nick and I quickly seized our chance, made our strong case, and within a month we had had a meeting with him in Whitehall (together with colleagues Mike Gerrard and

Dave Wardle of the EA), which was the beginning of him ultimately understanding the need for the Thames Tideway Tunnel and the inappropriateness of the alternatives (one of which was to attach plastic pipes to each CSO and run them on the riverbed to take waste water direct to the downriver WWTWs, a solution we dubbed the 'spaghetti' option).

At the 2011 Conservative Party Conference in Manchester, we had hired a room in the grand Midland Hotel to invite members of the party to visit and be updated about the project. I remember being in the room with Richard Aylard when the door opened and in came Victoria Borwick, Deputy Mayor of London. Seeing there was no one else in the room, she closed the very imposing door behind her. Victoria was in charge of Boris Johnson, Mayor of London's re-election campaign. The gist of the conversation was that she did not want anything that was being undertaken on the project, prior to the election, to potentially lose Boris any votes. We listened attentively.

Another of our more colourful meetings was not actually something organised by us. This followed a visit by the Chancellor, George Osborne, to China in early 2012, after which a high-powered Chinese Government delegation arrived in London to meet with the

Victoria Borwick, Deputy Mayor of London, with Andy Mitchell and me at the Conservative Party Conference in Birmingham, September 2014.

Receiving awards for *You Poo Too* short film at the tve Global Sustainability Awards 2014, at Bafta in London Piccadilly.

Computer-generated image of how Blackfriars Bridge Foreshore interception site will look when construction has been completed.

Sign on one of the underground access chambers to the Fleet Main Sewer, beneath the road access to Blackfriars Bridge.

View from an elevated platform looking up the Fleet Main Sewer. In the centre of the picture is the half-height weir that was constructed by Bazalgette.

Brickwork arch forming access chamber to the Fleet Main Sewer.

A heron on foreshore at Barnes, next to a plume of sewage discharged from a combined sewer overflow.

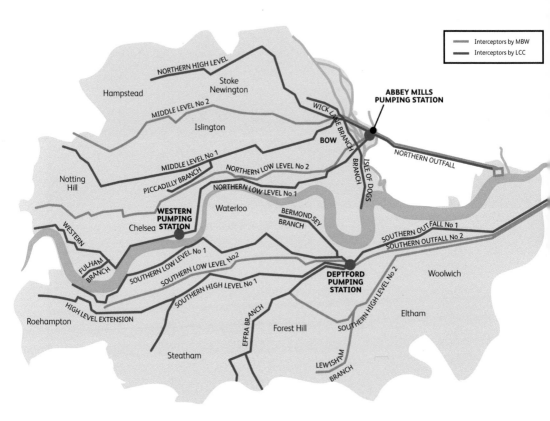

Map showing London's interceptor sewers, including main pumping stations.

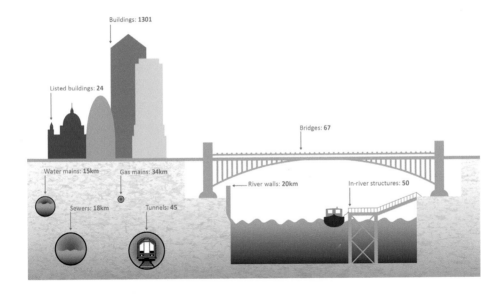

The existing infrastructure that will be crossed by the Thames Tideway Tunnel.

Computer-generated image of how the King Edward Memorial Park Foreshore site will look when construction has been completed.

Tideway Chief Executive Officer's rowing team during a charity event. The chimney of Bazalgette's Western Pumping Station can be seen on the left in the distance.

Combined sewer overflow in the southern abutment of Bazalette's Putney Bridge.

Montage of the River Thames from Tower Bridge. (Courtesy of Matthew Joseph)

Montage of the River Thames from Waterloo Bridge. (Courtesy of Matthew Joseph)

The Tideway Team outside the offices at The Point, Paddington in January 2015.

Speaking at the very successful Tideway Industry Day on the *Silver Sturgeon* in August 2013.

One of the two 8.8m diameter tunnel boring machines for the central section of tunnel (later named 'Rachel').

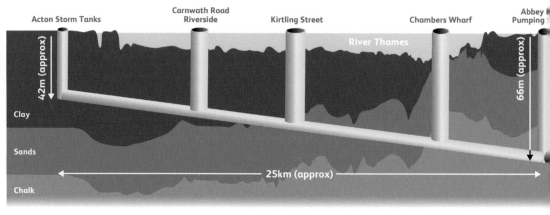

Longitudinal section of the Thames Tideway Tunnel showing the geology that each section will pass through.

Computer-generated image of how the Victoria Embankment Foreshore site will look when construction has been completed, as viewed from the embankment.

Computer-generated image of how the Victoria Embankment Foreshore site will look when construction has been completed, as viewed from the river.

Cross-section of Victoria Embankment Foreshore showing interception of low-level interceptor sewer, shaft and connection to main tunnel.

Stand-up paddle board yoga instructor Kristiana Thomas, near Kew Bridge. (Courtesy of Matthew Joseph)

Thames21 volunteer Jonothan Starkey on the Isle of Dogs, looking across to Greenwich Naval College. (Courtesy of Matthew Joseph)

Marine logistics expert Chris Livett at Butler's Wharf, with Tower Bridge in the background. (Courtesy of Matthew Joseph)

Tideway empolyee Stephen Shipley on the banks of Chiswick Eyot. (Courtesy of Matthew Joseph)

Tower Bridge driver Glen Ellis in the giant bascule chamber beneath the bridge's south tower. (Courtesy of Matthew Joseph)

Thames Rib Experience co-owner Charlotte Thompson, with the O2 Arena in the background. (Courtesy of Matthew Joseph)

First Mate on the Thames Clippers Peter Haining, looking across to the Emirates Air Line. (Courtesy of Matthew Joseph)

Thames21 Volunteer Jonothan Starkey on foreshore. (Courtesy of Matthew Joseph)

Nick Raynsford MP and Zac Goldsmith MP at a Thames Tideway Tunnel exhibition in the Houses of Parliament, June 2012.

Nick Raynsford pointing to the site in his Greenwich constituency at the Thames Tideway Tunnel exhibition in the Houses of Parliament, June 2012.

project management team. The delegation was responding to an invitation that the Chancellor had given to the Chinese Government to come and help the UK deliver its National Infrastructure Plan, of which the Thames Tideway Tunnel formed a prominent part. The Chinese delegation arrived in London holding a Memorandum of Understanding (MoU) for the finance and construction for the entire project, ready to sign! The very idea of forming an exclusive arrangement and abandoning planned competitions for construction and finance was unthinkable to the management team, committed as we were to delivering good value for money to customers; to say nothing of European procurement rules, which preclude any such exclusive bilateral deal-making. It took tact and time to avoid a rejection of the MoU by Thames Water becoming a diplomatic incident!

Such national-level (and in the latter case, international) engagement went up and down the political ladder, from the top with the Prime Ministers and those close to them, through key opposition Shadow Ministers, Select Committee and All Party Parliamentary Group Chairs and members, through to the constituency MPs who were so key to our being able to effectively engage with their constituents. While many of the latter were opposed to our plans (at least at first), Richard Aylard and Nick Tenant point out that one of our great successes was to maintain a broad cross-party consensus of support. This included Ian Pearson from Labour (both in Government and out); Richard Benyon from the Conservatives (again in post and out, together with his party colleague Zac Goldsmith, who supported our plans despite opposition from some of his own constituents); Nick Raynsford MP, who was also Chair of the All Party Parliamentary Infrastructure Group; and Baroness Sarah Ludford from the Liberal Democrats, who played such a key role in bringing the issue of pollution in the River Thames to the attention of the EU, UK Government and wider public.

One particularly interesting encounter was when two colleagues and I were summoned to appear before Margaret Hodge MP, Chair of the Public Accounts Committee, to answer questions about how we would ensure the project would deliver value for money for Thames Water's customers. The MP had a fierce reputation for calling business leaders to account. She was very challenging of how effectively we were managing expenditure and our confidence in the project outturn cost.

With regard to local Government, our engagements were targeted and extensive, with us taking every opportunity to present our case and listen to concerns. We had a difficult discussion with representatives of the London Borough of Wandsworth in March 2012 to outline our case and deal with concerns regarding the Barn Elms and Putney Embankment Foreshore sites, while similarly we met officials from the London Borough of Southwark in June 2013 to discuss issues regarding Chambers Wharf. It was also at local Government level that we came across some of our strongest opponents – not surprising, given that the issues were perhaps most keenly and personally felt at this level. Worthy opponents at this level included Stephen Greenhalgh, former leader of Hammersmith and Fulham Borough Council, who led a singing protest against us in Carnwath Road.

Meanwhile, in terms of NGOs, we have engaged throughout the entirety of our project with a variety of individuals from the Green movement and others often opposed to us, sometimes successfully and sometimes not. One of the principal criticisms we have faced in this regard is that the Thames Tideway Tunnel is not the greenest of solutions to London's current and future waste water requirements, though of course we argue that it is. Principal NGO opponents to this day include the Thames Blue Green Economy and Blue Green UK. The former describes itself as:

> a broad alliance including academics, civil engineers, economists, environmentalists, health practitioners, journalists, lawyers, politicians and scientists, all who are working together, to achieve the most cost effective and sustainable solutions for London's water-related environmental problems.

They particularly champion integrated water resource management (IWRM) for the Thames River Basin. The latter, meanwhile, are championing the Blue Green London Plan. They say they are campaigning for issues including (note these largely track the core concerns set out at the beginning of the chapter): linking urban water infrastructure (blue) to urban vegetation (green); reversing urban creep; creating greater climate change resilience; reducing flood and drought risk; encouraging the efficient use of water; reducing the heat island effect; lowering pollution level; and encouraging better biodiversity.

Both of these groups remain very vocal in their activities opposing the Thames Tideway Tunnel. However, we have also been successful in our engagement with other NGOs, such as the World Wildlife Fund, who were initially an opponent but are now much more supportive.

Media

Our engagement with the media was at a variety of levels: international, national and local level broadcast, print (both newspapers and trade publications) and social media. Their support or otherwise often reflected their audience's constituencies. For example, the *London Evening Standard* – perhaps the most important local publication, given it covered the whole geographic area of the capital – was a long-term opponent because of the disruption expected to be caused for its readership by the tunnel-building programme.

One of the lowest points was when the *Evening Standard* printed an article wrongly suggesting that the project would result in thousands of people being moved out of their homes. The turning point in their coverage, however, was when a journalist was given an expert-led tour of the sewers, where he was shown the environmental reasons behind the project's need (reflecting our broader strategy of ensuring that our most knowledgeable colleagues were always on hand to help with our communications outreach). From that point onwards, coverage in the *Evening Standard* has been broadly favourable. Similarly, other more local newspapers have also shown a keen interest in the progress of the Thames Tideway Tunnel if and when it affected their areas of coverage. Carnwath Road Riverside in Fulham and Chambers Wharf in Bermondsey have proved to be big draws in that regard.

A particular challenge we had was the negative articles that were printed in the London Borough of Hammersmith and Fulham's in-house newspaper, which the council used to vent their opposition to the project. The articles included a great deal of misinformation relating to noise, dust and light nuisance, as well as speculating about the health and wellbeing of residents and the sewage odours once the tunnel was operational. We were given no right of reply and had to use other communication channels to promote our key messages. Such vehicles for what was effectively one-sided propaganda were later banned by central Government.

On a more positive note, I undertook fifteen interviews over a number of years on a combination of BBC London and ITV London, where I was able to articulate the need for the project and the measures we would take to do all we could to minimise

BBC TV interview with Tom Edwards on the foreshore in Chelsea, adjacent to the Ranelagh combined sewer overflow in February 2012.

ITV TV interview with a copy of the DCO at the Institution of Civil Engineers, following the submission of the DCO in February 2013.

BBC TV interview with Tom Bateman in the Fleet Sewer, August 2013.

disruption to local communities and residents. These interviews had an enormous reach and helped raise the profile of the project in the capital.

At a national level, newspapers have only tended to cover our story at key times in the decision-taking process, and although some have questioned the need to spend such a large amount of money (the *Financial Times* at one point), broadly the coverage has been benign. For the broadcast media, again principally at local level, the interest in the Thames Tideway Tunnel has also been driven by interest during spikes in the consultation and planning application process. Understanding these paradigms has proved crucial in our ability to target our communications resources where they are most needed at any given time.

In order to better communicate the message of why the Thames Tideway Tunnel was needed, we commissioned a marketing company to make a short film. The title we decided upon was *You Poo Too*. It featured a young female actor who delivered a very convincing message about why the project was needed. The film was targeted at younger people and was a great success. Indeed, the film won the Creating a Better Environmental category and the Best Overall Film at the tve Global Sustainability Awards 2014 . It was a great night out and a chance to rub shoulders with some well-known celebrities.

The Thames Tunnel Now group at the Bazalgette Memorial on the Victoria Embankment with Sir Peter Bazalgette, November 2011.

The Thames Tideway Tunnel story has also broadly followed the rise of social media, and we have learned on the go how best to engage with stakeholders through its various channels, just like the rest of society. Twitter provides an excellent example of this, with Nick Tennant initially hosting the project's account under his own name – we quickly learned that this allowed opponents to personalise their opposition to the programme. Originally we also tried to respond to Twitter inquiries through the same medium as soon as possible, but this was almost impossible given the complexity of some of the issues. So today, we use a Twitter account just headed 'Tideway', and try to respond to detailed requests through other channels, such as email, where we can go into far more detail.

A particular ally we had when under fire from the media was the action group Thames Tunnel Now, a coalition of many environmental charities and community groups that represented a membership of 5 million people. On many occasions they issued press releases in support of the project, which were far more powerful than the same words issued by the project team itself.

We did all of our media activity against a backdrop of the broad public audience being unwilling to engage about the subject of waste water (unless directly affected by the project construction process). Waste water clearly doesn't make for compelling breakfast television!

Other Stakeholders

Outside of our engagement with the worlds of politics and the media, we have also engaged with a wide variety of other audiences to tell our story. Some of the most memorable for me were the various chances to brief members of the royal family. On 9 November 2016, I was very fortunate to spend an evening escorting into the event, and then sitting next to, the Princess Royal in the banqueting room of the Fishmongers' Hall at a livery dinner. She showed a keen interest in our story. This was the second time I had met her: the first had been in August 2013 at the Doggett's Coat and Badge Race (the oldest rowing race in the world, competed by apprentice watermen on the tidal River Thames) when we had had the chance to speak for thirty minutes. Meanwhile, the Duke of Edinburgh was similarly briefed by myself at the Doggett's Coat and Badge Race in 2014. Richard Aylard and I also briefed Prince Charles for over an hour and a half in Clarence House in a 2012 private update. He was very supportive, being particularly interested in how our activities might interface with the work the Prince's Trust

Martin Baggs and me briefing HRH The Princess Royal on the project at the Doggett's Coat and Badge Race, July 2013.

With HRH Prince Charles at the opening of the Lee Tunnel in September 2015.

was doing in Hammersmith and Fulham, and said it was important that as a project we created our own heritage.

When the Lee Tunnel construction was completed, Prince Charles toured London's deepest sewer. He also visited Abbey Mills Pumping Station: the Victorian system, which was opened by Charles's great-great-grandfather Edward VII while he was still Prince of Wales, was celebrating its 150th birthday.

We participated in as many industry-wide events as we could to ensure our case, and its key messages, was as well presented as possible. One of the many examples that sticks in the memory was the one-day Future Water 2013 event at the Royal Geographical Society on 25 June 2013, where we made our case about resilience in the water sector to an attentive audience from across our industry. In fact, I am the only person who has ever presented at seven consecutive UK tunnelling conferences – a good measure of how important it was for us to keep the market well informed about our progress! Further, a huge amount of effort was also undertaken from 2008 to 2014 in relation to similar market engagement abroad, including trips to many European countries, along with China, Singapore and New Zealand. The project is widely followed

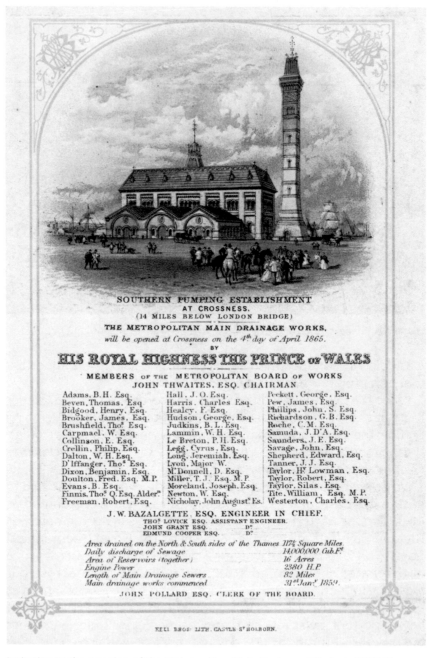

Invitation to the opening of the Metropolitan Main Drainage Works in 1865. The works was opened by the Prince of Wales at Crossness Outfall Works.

in Scandinavia because of the number of tunnels that have been and are being constructed.

It was also important for us to engage with local schools in our areas of activity, not only to reach out to the students, but also their parents (especially given the concerns of the latter regarding the impact of the construction process on their children). We therefore organised a variety of activities with local schools as part of our consultation, and later our engagement programme; for example, the visit organised to the Beckton WWTW for St Michael's School at Chambers Wharf on 6 December 2012.

Sewer visit for key stakeholders to Bazalgette's Hyde Park Sewer.

A visit with key stakeholders to the Fleet Main Sewer, standing adjacent to the original gearing mechanism for a penstock.

A visit with key stakeholders, including Mary Nightingale, ITV newsreader, and James Cleverly, Member of London Assembly, to Bazalgette's Hyde Park Sewer.

From early on in the project development, we organised sewer visits for key stakeholders and opinion formers. The vast majority of these took place in the Fleet Sewer. The rationale was to give individuals first-hand experience of the scale of the problem that needed to be solved and to show them how interwoven the sewerage system is in terms of London's underground infrastructure. The tours, which are akin to urban potholing, were very well received. Although many reluctantly agreed to attend, almost without exception people were full of enthusiasm after their visit. It was a unique experience: over the last 150 years, very few people had been given the opportunity to see a part of Bazalgette's masterpiece. Many key individuals attended, such as ITV News presenter Mary Nightingale.

Continuing Controversy

Another lesson the project team had to learn was that the communications campaign to build the Thames Tideway Tunnel never stops, it keeps going on and on, and will do so until the construction programme is completed. Those we have had to convince, or still have to, come from a wide variety of backgrounds – even Ofwat and Thames Water shareholders in the early days, the latter a very unusual experience for me. Another example of such opposition was the 2011 study commissioned by the London Boroughs of Richmond Upon Thames, Hammersmith and Fulham, Kensington and Chelsea, Southwark and Tower Hamlets, which looked at alternatives to the Thames Tideway Tunnel at the height of our engagement in the consultation phase.

Professor Chris Binnie, the TTSS's independent chairman, was still not convinced (sometimes he was more supportive, sometimes less), despite having led the body that initially recommended the storage and transfer tunnel solution to London's current and future waste water requirements. The *International Business Times* reported in 2015 that he remained concerned about its value for money, saying that since 2005 a number of other methods had come to light that would reduce the spill frequency and were technically viable and cost effective.[3]

Anecdotally, our most vocal high-profile opponents were dubbed within the team 'the four Bs': namely Professor Binnie, Lord Berkeley, Sir Ian Byatt and Martin Blaiklock. I remember a rather famous occasion at an industry event in the Houses of Parliament in 2012 (organised by Simon Hughes) when Professor Binnie was speaking out against the project. During his presentation he accidentally fell off the stage, quickly recovering to finish his speech, professionalism worthy of the West End stage. We had a great deal of correspondence with Sir Ian Byatt, who was one of our most outspoken opponents. He was the former Director General of Ofwat and the key architect of the UK water industry privatisation. He was particularly concerned about the need for the project and the cost for Thames Water customers. On one occasion, I visited him at home to debate our differences of opinion. However, he remains sceptical about the project to this day.

On 14 December 2011, water experts from across the capital gathered to debate the project at an event organised by the Chartered Institution of Water and Environment Management (CIWEM) in Belgrave Square. They debated a motion stating, 'This house considers that the Thames Tideway Tunnel would be worse value than controlling rainwater near source.' The motion was proposed by Professor Richard Ashley from Sheffield University and Lord Selborne (a Conservative Party Lord who has sat on the House of Lords Science and Technology Committee since 2005). Speaking against the motion were Richard Aylard and David Crawford, the Thames Tideway Tunnel team's chief engineer and resident international expert on CSOs. This debate concluded with a vote, with eight members of the audience supporting the motion, forty-eight opposing it and twenty abstaining, it therefore losing decisively. As Richard Aylard said afterwards: 'The overwhelming view from the floor was that it's not a question of having either the Thames Tideway Tunnel or SuDS, we need both.'

It was bizarre to continue the debate with Professor Richard Ashley at the Sustainable City Conference in Gothenburg in 2015. Whilst my presentation focused on the need for the project and the engineering proposals, Richard's was a talk on SuDS – yet it included many slides about why the Thames Tideway Tunnel was not needed!

TIMELINE

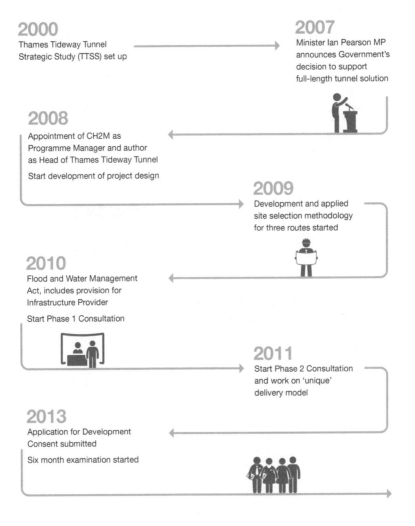

2000
Thames Tideway Tunnel
Strategic Study (TTSS) set up

2007
Minister Ian Pearson MP
announces Government's
decision to support
full-length tunnel solution

2008
Appointment of CH2M as
Programme Manager and author
as Head of Thames Tideway Tunnel

Start development of project design

2009
Development and applied
site selection methodology
for three routes started

2010
Flood and Water Management
Act, includes provision for
Infrastructure Provider

Start Phase 1 Consultation

2011
Start Phase 2 Consultation
and work on 'unique'
delivery model

2013
Application for Development
Consent submitted

Six month examination started

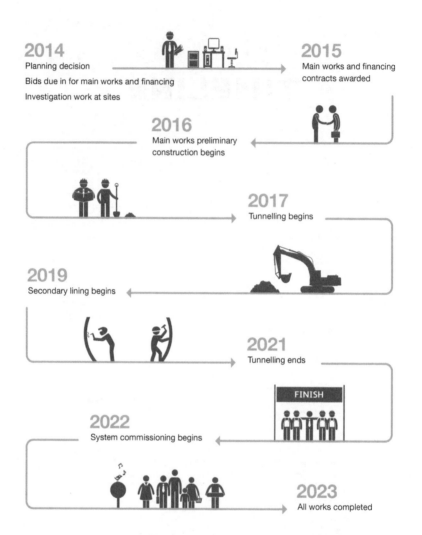

2014
Planning decision
Bids due in for main works and financing
Investigation work at sites

2015
Main works and financing contracts awarded

2016
Main works preliminary construction begins

2017
Tunnelling begins

2019
Secondary lining begins

2021
Tunnelling ends

2022
System commissioning begins

2023
All works completed

6

BAZALGETTE REBORN: A REMARKABLE ENGINEERING SOLUTION

London is not the first city to have to deal with a large volume of waste water, nor is the Thames Tideway Tunnel the first deep tunnel solution.

There are other examples of where 'deep tunnel' CSO projects have been constructed around the world to deal with the discharge of large volumes of waste water into watercourses. In June 2008, a small team of us visited Chicago, Milwaukee and Portland (Oregon) in the USA to speak to the teams who had delivered these major tunnelling projects. Without doubt the Portland scheme is the most similar to the combined Lee Tunnel and Thames Tideway Tunnel solution. The greatest benefit of the trip was inspecting these assets and gaining an understanding of the scale of such projects. There are other examples of deep tunnel solutions in a number of international cities, such as Paris, Vienna, Stockholm, Doha, Sydney, Washington DC. However, where London is different is that the Thames Tideway Tunnel is the first time that such a project will have been financed from the private sector.

The largest such project of its kind ever initiated, and due to be completed in early 2024, the 25km tunnel with a storage capacity of 1,240,000m³ is not only relevant for the engineering solution in its own right, but also what it will additionally deliver, both during its construction and also once operational. This includes not only the 95 per cent reduction in the amount of waste water being deposited in the River Thames and its tributaries, but also:

- The creation of 3 acres of new foreshore and measures to enhance the Thames Path.
- The ultimate creation of 4,000 direct, sustainable jobs, with a 1:50 apprentice and 1:100 ex-offenders employment ratio.
- In an environmental context the use of the river to transport or remove 90 per cent of the excavated material from the main tunnel drive sites (see below for detail).
- The engagement of a wide variety of local businesses and small and medium-sized enterprises (SME) for all aspects of the construction project.
- Our objective to transform health and safety performance on large infrastructure projects (an evolution from Thames Water's Triple Zero programme, which had committed us to zero incidents, zero harm and zero compromise).
- The introduction of leading lorry and cycling safety initiatives.
- The establishment of a training schedule for barge operators.

The Thames Tideway Tunnel is far more than just the physical entity of the infrastructure, and yet one can never get away from the fact that at the absolute core of the programme sits the engineering project itself.

It is important to note here the specific responsibilities of Thames Water and Tideway regarding the Thames Tideway Tunnel. This might seem to be an easy delineation to outline, but it is actually complicated by the fact that such responsibilities (both present and future) have evolved over time. Today, however, they are very clear. Tideway has responsibility to build and set to work the tunnel, and works at twenty-one sites (Thames Water being responsible for the other three), after which it will operationally be handed over to Thames Water. Tideway will then continue to exist, but as a much reduced organisation with specific responsibility to maintain the shafts and tunnels.

Engineering Overview

As detailed in Chapters 3 and 4, London's interceptor sewers, as designed by Joseph Bazalgette, were intended to cater for a population of 4 million people, and yet today they are having to cope with a figure well over 8 million and which is continuing to grow. Bazalgette's sewers are still in excellent working condition,

An integral part of the London sewerage system – a road gulley.

having been well maintained. The issue today is one of capacity. The system is very susceptible to rainfall events, given it utilises combined sewers that carry both the waste water and storm flow. Along with the other two components of the London Tideway Improvements (the upgrade of the five tidal WWTWs and the Lee Tunnel), the key part of the jigsaw puzzle is, of course, the Thames Tideway Tunnel.

This 7.2m diameter storage and transfer tunnel will run between 30m and 65m in depth, mainly below the River Thames. Starting in west London, the approved route for the main tunnel roughly follows the River Thames to Limehouse, where it then continues to the north-east under the Limehouse Cut, ultimately arriving at the Abbey Mills Pumping Station near Stratford in east London. There it will then be connected to the Lee Tunnel, which will transfer the waste water to the Beckton WWTW.

The detailed design and building of the project has been split geographically into three main works construction contracts, these being West (approximately 6,950m long), Central (divided into two, one section approximately 5,000m long and the second approximately 7,680m long), and East (approximately 5,530m long). Each section will be under construction at the same time, this to ensure that it can

7.2m diameter
Thames Tideway Tunnel

Three London double-decker buses would fit into the diameter of the Thames Tideway Tunnel.

be completed within the required timescales. To build the tunnel, the construction sites are spread across fourteen London local authorities, with the route passing under or close to seventy-five bridges, fifty in-river structures, 20km of river walls, 1,301 buildings, forty-five tunnels, 24km gas mains, 18km of sewers and 5km of water mains.

During the various stages of the planning phase, Thames Water carried out geotechnical investigations along the whole length of the proposed route of the main tunnel and long connection tunnels: the largest linear geotechnical investigation ever undertaken in London. This was designed to help to understand the type of ground conditions that the contractors would be tunnelling through, including London Clay in the western section, mixed sands and

148

gravels in the central section and chalk in the eastern section. Each of these types of geology requires a different type of Tunnel Boring Machine (TBM, see Appendix V for discussion of different types as appropriate for the different geologies featured), with the contractors being invited to examine the core samples before submitting any final bids so as to fully understand each area's specific tunnelling requirements.

The three main tunnel drive sites from which the TBMs will be driven are located at Carnwath Road Riverside in Fulham for the western section, Kirtling Street next to the Battersea Power Station development for the central section and Chambers Wharf in Bermondsey for the eastern section. The Frogmore connection tunnel will be driven from Dormay Street and the Greenwich connection tunnel will be driven from Greenwich Pumping Station.

The design outlined here was undertaken by Thames Water's project team. The key to ensuring the robustness of the tunnels through time (with a target of being fit for purpose for at least 120 years) was to ensure that its design took into account the aggressive environment it would encounter over this lengthy period. In that regard, the specific construction methods used for the Bazalgette interceptor sewers (which featured 161km of mainly cut and cover brick construction) were deemed inappropriate for a project of this scale. Therefore, the engineering design teams turned to the use of modern reinforced concrete, though even then it was deemed that modern design codes did not cater well for the envisaged challenges of the operational environment within the new system. A project was specifically undertaken to carry out a programme-specific durability assessment, this taking into account the unique conditions and properties of the new tunnel system; these results were then fed into the final design of the tunnel. About 20 per cent of design was completed before the main construction works tenders were issued. The remaining design is being undertaken by the main works contractors.

Many modern tunnels have either one or two reinforced concrete linings, but the linings serve different purposes for different tunnels. Our main tunnel and long connection tunnels will have two linings: a primary structural one to help the ground withstand water pressure and keep the tunnel watertight, and a secondary one to protect the primary lining segments and gaskets and provide additional durability. Both elements together will ensure a 120-year design life for the tunnel, with minimal need for maintenance.

The design process also included the use of modelling to design the various key components of the Thames Tideway Tunnel. Our expert in this regard at Tideway is David Crawford, the Chief Engineer of System Design and Operations. David says: 'the modelling process featured a series of different types of both computational and physical models depending on the desired outcome of the process.' In that regard, there were four types of model used. These were:

- Sewerage catchment models, used to determine how the system improvements will control CSO discharges and how variable the residual (controlled) discharges will be (given the variability of rainfall from year to year). Catchment models are also specifically designed to test how resilient the particular scheme will be to population and climate change (clearly very important with regard to London).
- Water quality models for the tidal River Thames, designed to determine the impacts of the polluting discharges on the ecology of the river, particularly from dissolved oxygen (and also to show how the Thames Tideway Tunnel would improve the condition of the river to meet the requirements set by the EA).
- Design models, which use computational fluid dynamics (CFD) methods and which were the first stage models for the design of the hydraulic structures in the system. These are complex computer models of the structures that show the flow regime (flow paths and velocities for example) through the structures in the system.
- Scale physical models of the hydraulic structures, either 1:15 or 1:10 in scale, to test how the system would perform under design flow conditions (and testing, for example, what impact closing penstocks would have on flow or how sediment passes through the structures)

Sites from which tunnels are driven (drive sites) are much larger in the construction phase than the sites where the tunnels arrive (reception sites). Therefore, determining the main tunnel's drive strategy was interlinked with site selection and involved the following steps:

- Grouping the shortlisted main tunnel sites into geographical zones that took account of the general location of sites and their proximity to each other.

- Reviewing how far apart the zones were and where the changes in geology occurred, as distance and geology influenced the feasibility of drive options.
- Assessing the types of sites that were in each zone, as that influences the availability of drive options, in order to arrive at all the possible options
- The five site selection disciplines (planning, environment, engineering, property and community) coming to a collective decision on the most suitable double drive, single drive and reception site in each zone.
- The five site selection disciplines working through comparisons of the drive options and their associated sites to reach the preferred drive strategy.

The tunnel drive strategy that resulted from this process features, for the western section (where the geology is largely composed of London Clay), a single-drive strategy using one TBM from east to west, specifically Carnwath Road Riverside to Acton Storm Tanks. The drive shaft for this TBM is reused as the reception shaft for one of the central drives.

For the central section (where the geology is largely composed of sands and gravels), a two-drive strategy with one TBM travelling east to west from Kirtling Street to Carnwath Road Riverside and another TBM driving west to east from (again) Kirtling Street to Chambers Wharf. The double drive shaft at Kirtling Street will support the two TBMs simultaneously.

For the eastern section (where the geology is largely composed of chalk), a single-drive strategy with a TBM running from Chambers Wharf to Abbey Mills Pumping Station. The Chambers Wharf shaft is reused as the reception shaft for the eastern-central TBM. To reduce scheduling risk and cost, the construction of the reception shaft at Abbey Mills was completed as part of the Lee Tunnel construction. This facilitates the quick connection of the TTT to the Lee Tunnel, thereby completing the full London Tideway Tunnels system.

TBMs will be used to excavate and primary line the main tunnel, which will then be followed by the installation of the secondary lining as a separate activity.

Only three CSO drop shafts will be built on the line of the main tunnel at Acton Storm Tanks, Blackfriars Bridge Foreshore and King Edward Memorial Park Foreshore. Therefore, connection tunnels are

Rory Stewart MP, Minister of the Environment, visiting the Lee Tunnel.

then required at a number of locations to specifically link the other CSO drop shafts with the main tunnel. The two long connection tunnels will also be constructed in the same way as the main tunnel using TBMs. The Frogmore and Greenwich connection tunnels link two and three on-line CSO drop shafts to the main tunnel respectively. We anticipate that the majority of the nine short connection tunnels, each linking one CSO drop shaft to the main tunnel, will be constructed using open-faced tunnelling techniques, including the application of sprayed concrete linings. However, in the case of the eastern section, such tunnelling techniques will be avoided where possible, given the difficulty of their use with regard to the prevalent chalk geology and the associated risk of high water pressure. On-line shafts on the Greenwich connection tunnel will be used in these locations to remove this risk.

In addition to the above challenges, the system design process has also had to contend with other significant issues. These included the need to dissipate energy and minimise air entrainment in the system, this being achieved through the use of sophisticated vortex drops. As Pump outlines,[1] such shafts are used to transport waste water from an overstressed existing sewerage system to other underground tunnels, in this case the Thames Tideway Tunnel. During the plunge of flow created by the interception of the CSOs, large volumes of waste water can fall significant distances. At the Fleet Main CSO, at Blackfriars Bridge Foreshore, up to 50 cubic metres per second (the weight of thirty-five average family cars per second) of waste water will fall 53m in depth. This creates two problems. Firstly, a huge amount of energy will be dissipated at the base of the drop shaft, which can over time destroy this base; and secondly, it means that air can be entrained in the flow. In order to overcome these problems, a vortex is designed that will 'spin' the flow into a spiral and create laminar (smooth) rather than turbulent flow. Initially, computational fluid dynamics was used to design the vortex for each drop shaft. Then a 1:50 scale model was built for each installation at Cranfield University. These physical models were constructed and then 'fine tuned' to produce the optimum design, which will lead to the least amount of air being entrained. If air is entrained in large tunnels it can cause major problems, as the air bubbles over time can congregate together and form air pockets. If these are located between shafts, storm flows entering the shafts and flowing towards the air pockets can lead to the vast forces involved compressing this air, very much like a spring, which will at a certain point recoil. The effect of this is that huge forces can then push back along the tunnels and send the flow up the shafts, and sometimes even through the manholes in the shaft cover. When this happened in Minnesota, USA, manhole covers lifted an articulated lorry up into the air.

The project featured on the BBC1 programme *Bang Goes the Theory* in April 2015, in an episode was entitled 'Can

Inspecting a hydraulic model of an interception shaft with David Crawford, at Cranfield University.

Filming for the BBC TV programme *Bang Goes The Theory* with Maggie Philbin in the Fleet Main Sewer, March 2013.

London's Sewers Cope?' Presenter Maggie Philbin and the film crew visited the Fleet Sewer to understand how Bazalgette's interceptor sewers still form the backbone of London's sewerage system today. They also visited Cranfield University to film the scaled hydraulic models. It was a very exciting experience, being involved, and gave the project exposure on prime-time evening viewing. I really admired Maggie Philbin for going deep beneath London's streets into the Fleet Sewer whilst having a fear of confined spaces.

Another key engineering issue was the ventilation and odour control of the system, particularly as many of the sites will operate close to residential developments. Blowers are employed at what we call active ventilation plants at three sites on the Thames Tideway Tunnel project (Acton Storm Tanks, Carnwath Road Riverside and Greenwich Pumping Station) and at two sites on the Lee Tunnel project (Abbey Mills Pumping Station and Beckton WWTW) to move air through the system to keep the atmosphere 'fresh' (meaning not septic). In addition, activated carbon in air treatment units at all sites will be employed to remove any odour from the airflow leaving the system. This will ensure nearby residents can't smell it. We are proud that we have minimised the use of complex equipment by employing passive pressure at most sites, hence negating the need for (much more expensive to buy and run) mechanical ventilation equipment.

Having been directly involved in the project since 2008, I can honestly say that planning, designing and now building such a large-scale construction project under the dynamic and living city that is London (one of the most famous examples of the built urban environment in the world) has called for innovation, particularly taking into account the challenges of London's varying geology and the aggressive waste water system environment, that has far exceeded anything I have experienced before.

The Engineering Solution in Detail

This is clearly a very complex programme and my aim here is to make what follows as accessible as possible to the general reader, while at the same time keeping available all of the relevant technical detail for those following my narrative who have an engineering background. To that end, I use as a key source the framework to describe the engineering solution as set out in the Thames Tideway Tunnel Specification Notice issued by DEFRA on 4 June 2014 (signed by Lord de Mauley, then Parliamentary Under-Secretary of State in the Department).

Construction

In terms of construction, the 'Thames Tideway Tunnel Project' is defined as the design, construction, testing and commissioning of the works prior to the acceptance date when Thames Water takes over operational control in early 2024. This includes:

- Having overall responsibility for the main works in its three sections and the system integrator.
- Having responsibility for the maintenance of the main works and the system integrator scope.
- Having responsibility for the operation of the Thames Tideway Tunnel infrastructure.
- Having ownership of the main works and system interceptor scope up to the point of acceptance by Thames Water.

Tideway has overall responsibility in terms of the construction of the following:

- All hydraulic systems and elements, including (but not limited to):
  o The interception chambers and connections with the existing sewerage system designed to intercept or influence the flow of waste water from the CSOs or the wider waste water network in accordance with the Operating Techniques (see below) and environmental permits.
  o The connection culverts designed to convey the flows from the interception chambers and connections to the drop shafts, chambers, and the vortex generators.
  o The valve chambers needed to house the penstocks and valves that control the flows and isolate the tunnels from the sewerage network.
  o The drop shafts and the vortex generators to drop flows down to the level of the tunnels.
  o The de-aeration and energy dissipation chambers.
  o The connection tunnels to convey the flows away from the drop shafts that are off the line of the main tunnel to the main tunnel.
  o The overflow chambers to discharge the flows to the tidal River Thames directly, or via other watercourse or the sewerage network when the main tunnel is not available to receive flows.
  o The shafts on the line of the tunnels for access or egress, and tunnel ventilation.
  o The main storage and transfer tunnel (and other tunnels) to enable the conveyance of waste water flow from Acton Storm Tanks to Abbey Mills Pumping Station, and into the Lee Tunnel for onward conveyance to the Beckton WWTW.
- The air management elements of the system, including (but not limited to):
  o The fans (at the active ventilation sites) for the ventilation and active management and treatment of air release from the Thames Tideway Tunnel infrastructure.
  o The air treatment plant to avoid nuisance odours from air removed from the Thames Tideway Tunnel infrastructure (at both active and passive ventilation sites).
  o The dampers to manage the air flow in and out of the Thames Tideway Tunnel infrastructure and to provide pressure relief.
  o The interception and overflow chamber ventilation system to allow air movement in and out of the chambers when operating.
  o The associated ducting and chambers, ventilation columns, acoustic enclosures and other ventilation structures.

- MEICA equipment, including (but not limited to):
  o The penstocks that will be provided to control the flow of waste water into the tunnels when operating, and to isolate the tunnels from the waste water when required.
  o The flap valves that will be provided to protect the sewerage network from inflow from the tunnels or the tidal River Thames.
  o The MEICA equipment associated with the air management system, penstocks and flap valves as described above.
  o The MEICA equipment associated with maintaining the performance of the existing sewerage infrastructure, including storm pump recirculation systems at the pumping station sites.
  o The instrumentation to monitor and record conditions within the drop shafts, interception chambers and the ventilation system.
  o The instrumentation used to enable the incumbent undertaker (Thames Water) to monitor and control the system (or its individual components) in accordance with the Operating Techniques (see below).
  o The instrumentation used to monitor and record flow discharge events as necessary to demonstrate compliance with the Operating Techniques.
  o The telecommunications works to allow the control and monitoring of the local sites and the system overall.
  o Broadly, any building services related to the construction and operation of the Thames Tideway Tunnel.
- Supervisory control and data acquisition (SCADA) works to enable monitoring and control of the overall system in accordance with the Operating Techniques and the environmental permits.
- Architectural elements of the overall system, including (but not limited to):
  o Buildings, lighting, boundary walls and fencing.
  o Hard landscaping and soft landscaping.
  o Structures (including ventilation columns), buildings and public realm landscaping works.
  o Kiosk structures to house the MEICA equipment.
  o The provision of all of the elements required for operational and maintenance access of the Thames Tideway Tunnel network.
- The following works will be undertaken by Thames Water:
  o Modifications within the existing sewerage network to influence CSO discharges.

- o Civil and MEICA equipment works at the Beckton WWTW including the Tideway Pumping Station, flow transfer system, inlet works and Tideway CSO outfall works including a bypass siphon tunnel.
- o Such other works and services that are incidental to or reasonably necessary to intercept and divert the combined waste water flow to ensure the interception, collection or influence of the CSO flows in accordance with the Operating Techniques and the environmental permits.

The design life for the Thames Tideway Tunnel project is 120 years, as set out in the previous section, meaning that it has to be resilient to changes that will affect the flows in the system, both from changes caused by the ever-increasing population of London and also, of course, from climate change. Resilience has therefore been a cornerstone of the infrastructure's design and construction.

Meanwhile, the route taken for the main tunnel heads south from the Acton Storm Tanks in west London to the River Thames, and then generally follows the route of the River Thames eastwards towards Tower Hamlets (see Appendix VI for the effects of hydrogeology in the project). At this point it branches to the north-east, away from the River Thames and under the Limehouse Cut to the Abbey Mills Pumping Station near Stratford. There the tunnel then connects to the Lee Tunnel, and the flows of waste water captured by the system will then be transferred onwards for treatment at Beckton WWTW. In specific detail, this route breaks down as follows, it having been finalised after the key stages of consultation exercise:

From the Acton Storm Tanks shaft, the main tunnel will run south through a developed urban area. The alignment then passes under gardens or under roads where feasible to avoid buildings as far as possible. It will then run under and along the route of the existing Acton Storm Water Outfall Sewer tunnel from Acton Storm Tanks to Netheravon Road.

The main tunnel alignment then crosses the above-ground District and Piccadilly Line railway near Stamford Brook Station, across Chiswick High Road, and then continues down Netheravon Road.

As the main tunnel then passes beneath the River Thames to the south side of the river, it crosses above the Thames Water Ring Main and then under the western playing fields of St Paul's School before turning eastwards parallel to Lillian Road. It will then continue eastward to the south side of St Paul's School's eastern playing fields, to the south of the Hammersmith Bridge (designed and built by Bazalgette) underneath residential properties in Riverview Gardens, and then back under the River Thames.

The Hammersmith connection tunnel will then join the main tunnel approximately 250m to the east of the Hammersmith Bridge abutment, opposite the park in front of St Edmund's Square. From here to the Carnwath Road Riverside shaft the main tunnel will run entirely under the River Thames, with two connection tunnels joining it from the Barn Elms drop shaft and the Putney Embankment Foreshore drop shaft. It will pass above the Thames Water Lee Valley Tunnel and beneath the Richmond to Fulham high pressure gas pipeline, with 8m of vertical separation. The internal diameter of the main tunnel changes from 6.5m to 7.2m at Carnwath Road Riverside. The Frogmore connection tunnel from the King George's Park and Dormay Street CSO drop shafts connects to the main tunnel at the Carnwath Road Riverside main tunnel shaft.

From Carnwath Road Riverside, the main tunnel will then generally follow the River Thames to Kirtling Street. To start with the alignment will then run east, crossing under the alignment of the Wimbledon to Kensal Green National Grid cable tunnel, approximately 110m east of the Carnwath Road Riverside main tunnel shaft.

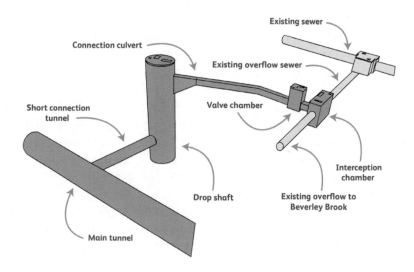

Construction graphic showing proposed interception at Barn Elms of the existing overflow at Beverley Brook.

East of Carnwath Road Riverside, the main tunnel will pass under an industrial estate including a PC World store and then pass through the northern span of the Wandsworth Bridge.

After crossing beneath Wandsworth Bridge, the main tunnel will then turn towards the southern bank to shorten the length of the Falconbrook connection tunnel.

Towards the Lots Road connection tunnel to the north, it will then pass below the second span (from the west) of the Battersea Rail Bridge before crossing back to the northern bank of the river to connect to the Lots Road connection tunnel.

From the Lots Road connection tunnel, the main tunnel will then follow the Chelsea Reach of the River Thames, passing under both Battersea Bridge (designed and built by Bazalgette) and Albert Bridge before reaching the Ranelagh connection tunnel.

From the junction with the Ranelagh connection tunnel, the main tunnel will then cross the mid-span of Chelsea Bridge and the second span (from the south) of Grosvenor Bridge.

From Grosvenor Bridge, the main tunnel will then turn toward Kirtling Street on the south side of the River Thames. The main tunnel will then pass beneath the jetty in front of the Cringle Dock refuse transfer station before reaching the Kirtling Street main tunnel shaft.

The main tunnel will then head south-east, staying to the north of a residential development recently constructed in order to minimise the impact on known current and future developments. The tunnel will then head east, passing under the jetty adjacent to the Tideway Industrial Estate as close to the south bank of the river as possible in order to minimise the length of the Heathwall/South West Storm Relief (SWSR) connection tunnel.

The main tunnel will then continue to the south of the river centreline and pass below the first span (from the south) of Vauxhall Bridge, in order to minimise the length of the Albert Embankment Foreshore connection tunnel and stay clear of existing buildings. It will then head north and cross the central span of Lambeth Bridge and beneath the Jubilee Line tunnels. As it approaches the Victoria Embankment Foreshore site, it will move to the western side of the river centreline, passing under the third span (from the west) of Westminster Bridge as close to the western side of the river as possible to reduce the length of the Victoria Embankment connection tunnel.

After the Victoria Embankment connection tunnel, the main tunnel will then pass below the second span (from the west) of Hungerford

Bridge and the BT St Martins General Post Office Tunnel. It will then continue on the northern side of the river centreline to pass under the second span (from the north) of Waterloo Bridge. Approaching the online Blackfriars Bridge Foreshore CSO drop shaft, it will then move to the northern side of the river with a straight section to take the tunnel alignment through the drop shaft.

The main tunnel will then pass below the middle of the second spans (from the north) of Blackfriars road and rail bridges. From the east of the Blackfriars bridges, the tunnel will follow the middle of the river as far as possible, passing below the middle spans of Millennium Bridge, Southwark Bridge, Cannon Street Bridge, London Bridge and Tower Bridge.

After crossing Tower Bridge mid-span, the alignment will then move across to the southern side of the river in front of St Saviour's Dock. This section of the route has been moved as far away as possible from the authorised navigation channel of the river to align with the main tunnel shaft at Chambers Wharf on the south bank of the river. The Greenwich connection tunnel from the Greenwich Pumping Station, Deptford Church Street and Earl Pumping Station CSO drop shafts connects to the main tunnel at Chambers Wharf main tunnel shaft. From Chambers Wharf the main tunnel will cross the river to the north bank to approach the King Edward Memorial Park Foreshore CSO drop shaft (scene of much of the heated debate covered in Chapter 5 regarding the consultation process), to take the main tunnel alignment through the drop shaft.

The main tunnel will then continue eastwards towards the entrance to the Limehouse Basin. The alignment will then turn northward to pass under the Old Sun Wharf as it cuts across towards the east side of the basin.

North of the basin, the main tunnel will thread between the tower block buildings of Basin Approach, to avoid passing directly beneath these structures. The main tunnel alignment also avoids here the high-rise buildings of Park Height Court on Wharf Lane and The Mission on Commercial Road. It will then follow the Limehouse Cut towards the Abbey Mills Pumping Station.

Following the Limehouse Cut as far as the Blackwall Tunnel Northern Approach Road, the main tunnel will then pass under the low-rise buildings at the Barratt Industrial Park as it turns to a more northerly direction where it crosses the River Lee. Keeping to the west of the gas holders, the main tunnel will then cross under the surface rail tracks of the District Line and across the Channelsea

River, passing onto the Abbey Mills Pumping Station land to connect to the Lee Tunnel at the main tunnel reception shaft.

As can be imagined given the scale of this route, geology is a significant factor in its construction. The main tunnel travels through London Clay in the west, mixed sands and gravels (Lambeth beds and Thanet sand) in the central region and then chalk in the east, with each of these requiring a different TBM given the different ground conditions in which they will operate. Overall, four TBMs are being used for the main tunnel, together with two others for the two long connection tunnels. The TBMs are the most vital component of the whole tunnel construction process, each comprising a cutterhead which actually performs the engagement cutting process with the strata as the tunnel advances, a ventilation system, and finally a conveyor or pump system to remove the excavated material (plus a control room to guide the TBM), with the tunnel segments then being placed in position as the cutter head advances.

Building the Thames Tideway Tunnel infrastructure will require two specific types of construction sites: main tunnel sites where the TBMs will be launched or received (main drive sites/tunnel reception sites), and also CSO interception sites where the existing CSOs will need to be connected to the new Thames Tideway Tunnel. The latter connections will be facilitated by further new and short connection tunnels/culverts, which are needed to transfer the flows of waste water from the existing sewerage system to the new tunnel.

All twenty-four sites are under the aegis of Tideway, except for three that remain under the direct control of Thames Water, these being sites where the existing waste water infrastructure in conjunction with other necessary CSO control work needs to be undertaken. The twenty-four sites are broken down into six specific types: main tunnel drives, main tunnel receptions, CSO interceptions, long connection tunnel drives, short connection tunnel drive/receptions, and existing modification sites.

Site	Type
Acton Storm Tanks	Main tunnel reception site and CSO interception site.
Hammersmith Pumping Station	Short connection tunnel drive site and CSO interception site.

Barn Elms	Short connection tunnel drive site and CSO interception site.
Putney Embankment Foreshore	Short connection tunnel drive site and CSO interception site.
Dormay Street	Long connection tunnel drive site and CSO interception site.
King George's Park	Long connection tunnel, reception site and CSO interception site.
Carnwath Road Riverside	Main tunnel drive site and main tunnel reception site and long connection tunnel reception site.
Falconbrook Pumping Station	Short connection tunnel drive site and CSO interception site.
Cremone Wharf Depot	Short connection tunnel drive site and CSO interception site.
Chelsea Embankment Foreshore	Short connection tunnel drive site and CSO interception site.
Kirtling Street	Double main tunnel drive site.
Heathwall Pumping Station	Short connection tunnel drive site and CSO interception site.
Albert Embankment Foreshore	Short connection tunnel drive site and CSO interception site.
Victoria Embankment Foreshore	Short connection tunnel drive site and CSO interception site.
Blackfriars Bridge Foreshore	CSO interception site.
Shad Thames Pumping Station	System modification site (an operational pumping station within the existing London sewerage network, so the responsibility of Thames Water).
Chambers Wharf	Main tunnel drive site and main tunnel reception site and long connection tunnel reception site.
Earl Pumping Station	CSO interception site.
Deptford Church Street	CSO interception site.

Greenwich Pumping Station	Long connection tunnel drive site and CSO interception site.
King Edward Memorial Park Foreshore	CSO interception site.
Bekesbourne Street	System modification site (an operational sewer within the existing London sewerage network, so the responsibility of Thames Water).
Abbey Mills Pumping Station	Main tunnel reception site.
Beckton Sewage Treatment Works	WWTW modifications (an operational WWTW, so the responsibility of Thames Water).

There are numerous designs for the CSO interception sites because construction of their drop shafts varies depending on the required depth, on the amount of flow the shafts need to carry and on the geology. The drop shafts come in the form of concrete cylinders, each of which has an internal diameter of between 6m and 22m, and being between 20m and 60m deep. Also needed at the CSO interception sites are the ventilation structures to allow air to circulate in and out of the shafts, with construction of the overall shaft infrastructure at the sites requiring between three and four years to complete. Seven of the CSO interception sites will be constructed out into the foreshore, this will create an additional 3 acres of new public realm. One of our key commitments here regarding the overall Thames Tideway Tunnel construction programme is that, once complete, each site will be landscaped to fit in with the local environment, minimising any long-term impact from the construction process.

Meanwhile, at the main tunnel drive sites/main tunnel reception sites, four principal activities are required. These are: vertical shaft construction, in the form of a concrete cylinder 25m to 30m in diameter and between 40m and 60m deep; preparing the site for the TBM; TBM assembly and lowering into the shaft; driving the TBM to excavate and line the main tunnel.

Once the boring process begins, as the TBM moves forward precast concrete segments will then be brought in and lifted into place and bolted together, this creating the tunnel structural lining in a process designed to operate twenty-four hours per day. The excavated material is then transported out of the tunnel via the conveyor belt or pump system, after which it is processed before being taken off site. One factor to bear in mind here is that in some of the sections being tunnelled through by the TBMs the soft rock is fissured, through which water can pass (see Appendix V for the specifics of the relevant geological strata). Therefore, in these regions, particularly at 65m underground, the workface has to be protected from up to 7 bar of water pressure.

Back to the excavated material, of which there will be approximately 5 million tonnes. To minimise disruption to London's roads, the project team has committed to use the river as much as possible to transport materials both in and out of the construction sites. This was a key part of the DCO for the Thames Tideway Tunnel when announced on 12 September 2014, where we are committed to removing 56 per cent of all materials by river where viable (the idea again clearly being to minimise road traffic disruption). For this marine transport function, four main methods of contracting the marine transportation were considered:

- A single marine operator contracted directly to Tideway.
- Each of the main works contractors to be fully responsible for their own marine operations and disposal.
- The contractor for the eastern section to be responsible for the transhipment and disposal of all excavated material removed by river, they being closer to the final receptor site. The contractors for western and central sections would then be responsible for the marine operations from their sites to the transhipment site or receptor site in the eastern section.
- The contractor for the eastern section being responsible for taking all of their own excavated materials directly to the final receptor site, with the contractor for the central section tasked with managing the trans-shipment site(s) and taking all remaining excavated materials to the final receptor site(s), and the contractor for the western section tasked with taking all of their excavated material to the transhipment sites for the central contractor to manage thereafter.

After careful consideration the fourth option was chosen, that being the one in operation today.

Operation

The Thames Tideway Tunnel project will move to the operational phase from the end of the system commissioning period (late 2022). The operation of all the new infrastructure will be by Thames Water, who will be responsible for ensuring that the system's shafts and the tunnels provide a minimum of 1.24 million cubic metres of volume so that the infrastructure can be used as planned, in conjunction with the wider sewerage network, to actively manage and influence flows from the CSOs in accordance with the Operating Techniques and the environmental permits. The maintenance of the shafts and tunnels will be undertaken by Tideway. The maintenance of the remaining infrastructure will be handled by Thames Water.

The engineering solution selected for the Thames Tideway Tunnel has been designed with the goal of minimising operational intervention and maintenance costs throughout the 120-year lifecycle once it becomes operational in early 2024. For this reason, the operational systems have been designed to be as simple as possible, using gravity to transfer the flows from the Acton Storm Tanks in the west to the Abbey Mills Pumping Station at the eastern limit of the Thames Tideway Tunnel system. Monitoring equipment to control the storage and transfer of flows in the tunnels will be supervised at a centralised location, along with a control and automation system for the isolation penstocks at each of the CSO connections between the existing sewer network and the Thames Tideway Tunnel system.

Once operational, the waste water diverted into the Thames Tideway Tunnel will be stored in the tunnel system and then, at the optimal moment, pumped out for treatment at the Beckton WWTW in east London. Storage is the key here. If the contents of the tunnels were released as they arrived at the pumping station during an extreme weather event, then the WWTW would be overwhelmed. If all the CSOs discharged their maximum flow at the same time (an extremely unlikely scenario), the flow would peak at 430 cubic metres/sec. The capacity at Beckton WWTW is 27 cubic metres/sec. Further, to accommodate the flows from the London Tideway Tunnels, six pumps are required at the pumping station at Beckton WWTW.

A typical Thames Tideway Tunnel CSO interception site, once constructed, will feature:

- An interception chamber to intercept the flows en route to the outfall, and then divert them via other structures to the main tunnel. When the tunnel system is full, the chamber would revert to existing conditions and allow flows to discharge into the river.
- Flow control equipment and penstocks incorporated into the interception chamber where practicable (though many will need to be housed in separate valve chambers).
- Ventilation structures, such that when the tunnel system fills with waste water flows, the air in the system can be displaced, and when the flow is pumped out then the air can be returned. The design also incorporates a means of refreshing the air within the system when it is empty. The air management system involves a combination of air extraction structures, air intake structures and air treatment equipment.
- An electrical and control kiosk, featuring the electrical or hydraulic components of the flow control equipment and air treatment plant.
- The connection culvert, transferring flows from the interception chamber/valve chamber into a drop shaft. The culvert can comprise several sections, depending on the location of the valves and penstocks. The longest connection culvert is at Blackfriars Bridge Foreshore.
- The CSO drop shaft, transferring flows vertically from the connection culvert to the connection tunnel/main tunnel via a vortex drop structure inside the drop shaft. The vortex drop structure dissipates the energy created by the large drop height and ensures air is not entrained in the flows at the bottom of the drop shaft.

The connection tunnel, transferring flows from the CSO drop shaft to the main tunnel via one of three basic arrangements, these being: a connection tunnel between a drop shaft and a shaft on the main tunnel; a connection tunnel between a drop shaft and the main tunnel; a connection tunnel between a drop shaft and another drop shaft. Connection tunnels are not required when the drop shaft is located on the main tunnel.

At three CSO interception sites, additional works are being undertaken to connect to the Northern Low Level Sewer No. 1 and divert some of its flow during rainfall events into the drop shaft and the main tunnel. This will avoid the need for interception sites at ten of the CSOs that the Thames Tideway Tunnel project must control, because their flows would be handled by the capacity created in the Northern Low Level Sewer No. 1. These ten CSOs are located along the northern bank of the River Thames (Chelsea Embankment and Victoria Embankment). Intercepting them would entail significant additional works on or near the embankment and in the foreshore, of a similar scale to the works proposed at Victoria Embankment Foreshore. Therefore, the connection to the Northern Low Level Sewer No. 1 at the three sites (where works were required anyway) significantly reduces the project's impact on road traffic and the marine environment. It also reduces project cost and makes the project more sustainable.

Meanwhile, once constructed, the main tunnel will feature:

- The main tunnel shafts. Here, where CSO flows enter the tunnel system at a main tunnel shaft, these act as drop shafts (e.g. at Acton Storm Tanks), while at other main tunnel shafts the flows would pass through the bottom of the shaft (e.g. at Kirtling Street).

Computer-generated image of how Chelsea Embankment Foreshore will look after construction has been completed.

- Ventilation structures, where the air management system involves a combination of air extraction structures, air intake structures and air treatment equipment.
- The electrical and control kiosk, housing the electrical or hydraulic components of the air treatment plant. Where CSO flows enter the tunnel system at the main tunnel shafts, the electrical or hydraulic components of the flow control equipment would also be housed in an electrical and control kiosk.

Further, the two CSOs that would be controlled by modifications to the existing sewerage system rather than interception require worksites, being those at Shad Thames Pumping Station and Bekesbourne Street (which discharges via the Holloway Storm Relief CSO), located under an apartment previously owned by Petula Clark. These works will be delivered by Thames Water.

The Contracting Process for the Construction Programme

While it is fair to say that the estimated cost for the Thames Tideway Tunnel scheme has increased steadily since it was first proposed in the early 2000s as part of the TTSS work, this to my mind only reflects the development of the project and the greater understanding and knowledge of the component parts that are required to successfully deliver the scheme. The estimate stated in 2004 at the time of the TTSS report announcement was £1.7 billion, including the WWTW upgrades and the Lee Tunnel. At this time very little design work or site investigations had taken place and there was only a very limited understanding of how the tunnel would be constructed or how the thirty-four CSOs would be intercepted or controlled. The estimate was a 'top down' evaluation based on unit rates for tunnelling and the other main construction activities. To put this in perspective, it was undertaken by a couple of engineers over a period of a few weeks.

Following the appointment of CH2M Hill in March 2008, the design of the project was developed and activities such as the site selection and tunnelling strategy were undertaken. This led to the production of the outline design that was used as the basis for the first phase of consultation in September 2010. With a far better understanding of the scope of the project, a 'bottom up' estimate was undertaken, which priced all the component parts of the scheme.

At this time the estimate was revised to £3.6 billion*, a sum which also included a far better appreciation of the land costs for the sites required to construct the Thames Tideway Tunnel. Following feedback, particularly from the first phase of consultation, the design of the project was changed significantly. Two of the main tunnel drive sites changed and many of the CSO interception sites were dramatically amended. Through the two phases of consultation the overall design of the project, in relation to tunnelling strategy and the general arrangements of the construction sites, was becoming far more 'fixed'. A revised estimate in 2011 of £4.2 billion was produced to reflect the revised scope that was now far better defined. By this time we had approximately 200 people working on the project and the estimate involved dozens of team members over a number of months. Despite the inflation that occurred since this time, the cost estimate for the project has remained static at £4.2 billion. This cost stability is very rare during the development of a major infrastructure project. The National Audit Office report entitled 'Review of the Thames Tideway Tunnel' (published in March 2017) stated that 'Cost estimates have risen over time during planning, but have remained relatively stable since 2011.'

As detailed in Chapter 4, following detailed analysis it was decided that the best means to proceed with the Thames Tideway Tunnel was to appoint a regulated (through Ofwat) IP to provide best value for money, with Tideway being selected as the preferred bidder on 13 July 2015. However, procurement of the main works and system integrator contractors was actually initiated earlier, in 2013 by Thames Water, and completed by Tideway as the IP in 2015.

It is now clear to all at Tideway that a well-planned, structured and clearly focused engagement by Thames Water with the key market players in this sector, in advance of the contractor procurement process, helped us design a packaging strategy and a form of contract that promoted the interest of the Thames Tideway Tunnel project. Additionally, it appealed to the market (wide interest in bidding for this huge and complicated contract being vital to ensuring value for money). There were also clear benefits realised during the procurement process, in terms of the good lines of communication having already been established, with

---

* In 2009 prices.

each tenderer during the extensive pre-bid market engagement activity already having a pre-existing open dialogue with Thames Water. This helped us later to fully understand the dynamics within each tenderer organisation, and to predict accurately the tenderers' reactions to major events during the procurement process. Overall, this helped facilitate the process of generating and maintaining genuine competitive tension in the procurement process.

This contractor procurement strategy also benefitted greatly from the experiences of best practice that were developed in this regard for other recent similar projects in London, such as the London 2012 Olympics. Thus we were able to set out within our wider contractor procurement strategy a robust framework for the management of the process. The preceding analysis undertaken in the development of this procurement strategy also informed the development of the evaluation criteria for the bids, as well as help-ing with the drafting of the main works contracts (for the three sections of construction along the line of the tunnel) and system integrator contractor, thereby ensuring there was a single consist-ent 'golden thread' running through the entire process. Further, the decision to split the main works tunnelling contract into the three parts, and also the associated adoption of the Win Only One Rule (WOOR), by which each winning bidder could only win the con-tract to build one of the three sections, were critical decisions that had far-reaching consequences for the overall procurement and the wider Thames Tideway Tunnel project. To date, as we acceler-ate the construction process, this decision appears to have brought significant advantages for the project in terms of bringing greater commercial resilience to the programme.

In terms of the scope of the work regarding the three sections of the main works that the bidders would be actually bidding to win, in summary these were: the design works; the enabling works; the temporary works; the construction works; the tunnelling and excavated material removal; the CSO interception works, the marine transportation and logistics; the ground stabilisation and treatment works; the utilities diversions; MEICA activities; SCADA activities; the structural engineering works; the civil engineering works; the architectural landscaping works; the protection of third party assets; the commissioning activities; and such other works and/or services as were necessary to deliver the Thames Tideway Tunnel project, importantly including (but not limited to) obtaining and maintaining all necessary consents, undertaking works and services to implement the agreed legacy strategy, and crucially, stakeholder management.

Additionally, it was also considered critical that certain elements of the wider construction work should be carried out in the same way across the whole project (rather than being split into three) by a single contractor. These two elements were access control and security, and occupational health. These two key contracts across the whole of the construction process were procured by Thames Water under Framework Agreements with the intention that the contractors would act as key subcontractors to each of the three main works contractors.

On 29 July 2013, Thames Water announced that a contract notice for work for the Thames Tideway Tunnel had been published in the *Official Journal of the European Union*. Then, following the issuing and receipt of completed PQQs, the water utility invited shortlisted contractors to tender between November that year and April 2014. The successful preferred tendering contractors for the three main works contracts were announced in February 2015:

West section: a joint venture between BAM Nuttall, Morgan Sindall and Balfour Beatty Group, initial contract cost £416 million. This consortium beat others including Costain with Vinci Construction Grands and Bachy Solentanche, another with Dragados and Samsung, and finally one featuring Ferrovial Agroman UK with Laing O'Rourke Construction.

Central section: a joint venture between Ferrovial Agroman UK and Laing O'Rourke Construction, initial contract cost £746 million. This consortium beat others including BAM Nuttall, Morgan Sindall and Balfour Beatty Group, another with Costain with Vinci Construction Grands and Bachy Solentanche and finally Skanska with Bilfinger, and Razel Bec.

East section: a joint venture between Costain, Vinci Construction Grands Projects and Bachy Soletanche, initial contract cost £605 million. This consortium beat others including one featuring (again) BAM Nuttall, Morgan Sindall and Balfour Beatty Group, another with Bechtel with Strabag, a further one with Bouygues Travaux Publics and finally one with Hochtief and Murphy.

At the time of the selection, Andy Mitchell, chief executive at Tideway, said:

We have selected our preferred bidders to work on the three Main Works packages because we have absolute faith in their ability to carry out these major pieces of work safely, considerately and sustainably and we are looking forward to working with them to offer the thousands of jobs that will help make this project a reality. This is not just an engineering project, this is about reconnecting London with its river by cleaning it up and making it something that is integral to our city, for the growing population, thriving businesses and to increase leisure uses. This is a unique opportunity to be involved in improving London's environment and economy and we're very excited for what the future holds.

The award of contracts to the three preferred tendering contractors was officially announced on 24 August 2015, in time for construction on site to begin in January 2016.

The Operating Techniques for the Thames Tideway Tunnel

The Operating Techniques (an operating agreement between Thames Water, and later Tideway, with the EA) through which the London Tideway Tunnels and their interaction with the Beckton and Crossness WWTWs will be operated, are written large both during the current build process and also once the system is operational. As set out by the Thames Tideway Tunnel Specification Notice issued by DEFRA, they cover both the operation and maintenance of the programme. The maintenance of the works and the operation of the Thames Tideway Tunnel Infrastructure, subject to any agreements between the IP and the incumbent undertaker (to recap, Thames Water) or third parties are included prior to the acceptance date of the completed tunnel in early 2024. The operation and maintenance of the IP Owned Structures (the shafts and tunnels and supporting infrastructure), the periodical inspection of the IP Owned Structures, the routine, planned and reactive maintenance activities for the IP Owned Structures and allowing the transfer of combined sewage through the Thames Tideway Tunnel Infrastructure are all included from the acceptance date.

The Operating Techniques are the culmination of many years of engagement between Thames Water (and now Tideway) and the EA, with the aim of scoping the project that will be determined by the

EA to be capable of meeting the requirements of the part of the EU's UWWTD that deal with limiting pollution from waste water overflows through CSOs. This engagement between Thames Water and the EA stretches back to the TTSS, which set the following objectives as the means of physically securing compliance with the UWWTD:

- To limit ecological damage in the River Thames and its tributaries by complying with the dissolved oxygen standards developed for the tidal River Thames by the TTSS.
- To help protect all river users by substantially reducing the elevated health risk due to discharges through CSOs.
- To reduce the frequency of operation and limit pollution from those CSO discharges that cause the most significant aesthetic pollution, to the point where they will cease to have any significant adverse effect on the river and its tributaries.

As set out above, part of the TTSS study identified the thirty-six CSOs that were unsatisfactory in terms of frequency of discharge and environmental impact. The thirty-six CSOs were selected as being the ones that in a typical year discharged flow that caused a material level of pollution to the tidal River Thames. The Thames Tideway Tunnel will intercept and control thirty-four CSOs. The two remaining will be dealt with by the Lee Tunnel and a small Thames Water project at Wick Lane.

At the same time, work by Thames Water (reviewed by the EA at the time) also included extensive modelling of the sewerage network and river water quality to determine the most cost-effective size of tunnel that would achieve the objectives set by the TTSS (this process described above in Chapter 4, the tunnel ultimately being 7.2m diameter for the most part), and in particular to achieve the dissolved oxygen targets and to limit the number of discharges from the thirty-six unsatisfactory CSOs to no more than three to four per annum. This was then fed into the development of the Operating Techniques for the London Tideway Tunnels, the ultimate set of rules that govern: the opening and closing of the penstocks that change the flow at the CSOs from either entering the tunnel system or being diverted into the River Thames or its tributaries (or in the case of the Lee Tunnel, from entering the River Lee via Channelsea Creek at Abbey Mills Pumping Station); and the switching on and off of pumps controlling flows at Abbey Mills Pumping Station, Greenwich Pumping Station and the Beckton WWTW.

Specifically, the operational regime of the Operating Techniques during wet weather events for the London Tideway Tunnels is as follows (the level of detail is included here for the specialist reader or researcher):

- When the waste water level in the tunnel system is below a specified level (-26m Above Ordnance Datum, AOD), the tunnel penstocks are open, diverting CSO flows into the tunnel.
- During heavy rainfall, once the waste water level in the tunnel system reaches -26m AOD, the tunnel penstocks are closed and CSO flows are diverted into the River Thames or its tributaries. The flow continues to enter the tunnel system at Abbey Mills Pumping Station.
- Once the waste water level in the tunnel reaches -20m AOD, the bypass pumps in the Lee Tunnel pump out shaft at Beckton WWTW will start to pump the contents of the tunnel system to the Tideway CSO at Beckton WWTW, discharging into the River Thames.
- Once the waste water level in the tunnel reaches -3m AOD, the penstock at Abbey Mills is closed and CSO flows at Abbey Mills Pumping Station are diverted to the River Lee/Channelsea Creek.
- As the rainfall intensity reduces or stops, and the waste water level in the tunnel system drops to -10m AOD, the penstock at Abbey Mills Pumping Station is opened, allowing residual CSO flows at Abbey Mills Pumping Station to be diverted to the tunnel.
- Once the water level drops to -24m AOD, the bypass pumping at Beckton WWTW will stop.
- Once the water level drops to -41m AOD, the tunnel penstocks are reopened.

In extreme wet weather the tunnel penstocks will actually be closed at -30m AOD, and bypass pumping will start at -28m AOD. This is designed to recognise the quicker rate the tunnel system fills during high-intensity rainfall events.

The Operating Techniques for the London Tideway Tunnels also specify the circumstances in which the tunnel system is pumped out into the inlet works at the Beckton WWTW. This is a function of the flow arriving at the waste water treatment works via the Northern Outfall Sewer and also the level of flow at the inlet works. The overall philosophy underpinning the Operating Techniques is designed to ensure that the CSO at Abbey Mills Pumping Station is capable of discharging to the tunnel system for as long as possible, beyond when the tunnel has been closed to CSOs along the River Thames. The reason for this is because the lower Lee Valley and River Lee have benefited from the post-2012 Olympics legacy, and are already a focus for extensive urban regeneration. Therefore, by the time the Thames Tideway Tunnel is operational, its London Tideway Tunnels partner the Lee Tunnel will have been operational for eight years, and greatly contributing to the environmental improvements in the Lee Valley. The control arrangements incorporated in the Operating Techniques have further been established to limit the discharges at Abbey Mills Pumping Station to a one-in-ten-years return period (this compared to the three to four discharges per year at the other CSOs connected to the tunnel system).

The control of the tunnel system will be completely automated, with the facilitating valves opening and closing in response to signals provided by waste water level sensors and the pumps starting and stopping in response to signals sent from flow meters and waste water level sensors. In the specific case of the Thames Tideway Tunnel, this instrumentation, control and automation system will be provided by the system integrator, a fourth and separate contract to the main works contractors, it then being handed over in an operational context to Thames Water once operational (who already have control of the function regarding the Lee Tunnel). The only manual intervention will be by the operating authority, Thames Water, when required to change the system mode from the 'Wet Weather' mode to the 'Extreme Weather' mode (in response to weather forecasts).

An Interface Agreement and the Operations and Maintenance Agreement provide a template by which Tideway and Thames Water (with responsibility for the Lee Tunnel, and also later the operation of the Thames Tideway Tunnel once built and handed over) facilitates the Operating Techniques.

As the design authority, Tideway has specific obligations under the Interface Agreement to ensure that, firstly, the designs for the Tideway works and the Thames Water works integrate with each other and the sewerage system such that the two London Tideway Tunnels can be operated in accordance with the Operating Techniques. Secondly, Tideway must ensure that the Tideway works when completed will be fit for purpose, and that the London Tideway Tunnels can be operated in accordance with the Operating Techniques.

Tideway, as the design authority, and Thames Water, as the operating authority, also have an obligation under the Interface Agreement to cooperate with each other in order to optimise the operation of the Thames Tideway Tunnel with the sewerage system (including the Lee Tunnel) in accordance with the Operating Techniques, this in consultation with the EA.

As the design authority, Tideway has an obligation under the Operations and Maintenance Agreement to maintain the Tideway-owned structures so that they are available for use and are capable of being operated in accordance with the Operating Techniques. Meanwhile, as the operating authority, Thames Water also has an obligation under the Operations and Maintenance Agreement to operate and maintain the wider sewerage system (which includes both of the London Tideway Tunnels) so as to enable its operation in compliance with the Operating Techniques.

This might all seem complicated to the reader, but it's not in reality: the Operating Techniques are simply the way Tideway and Thames Water are building and will operate the two London Tideway Tunnels.

In the entire time that we have been developing the project, we have reached out to other major projects to seek their advice. We have been particularly keen to learn the lessons from London 2012 Olympics and Crossrail. As an example, for many years I had four-monthly meetings with Chris Dulake, Chief Engineer of Crossrail, and John Trounson, Project Director, of National Grid Tunnels, where we shared project information. I would particularly focus on what could have gone better on these major tunnelling projects and what either Crossrail or National Grid would have done differently had they had their time again. These meetings directly influenced our approach to the project, procurement of main works contractors and the development of the Operating Techniques. For example, we procured the design and build elements of the project into single packages, for the reason given previously. We also specified that there would be a maximum of ten-hour shifts for underground construction workers (no maximum period was specified in the Crossrail contracts). Finally, we required a higher proportion of contract staff to be employed directly through PAYE, rather than hired through sub-contractors. To reinforce these initial meetings, for two years from 2012 to 2014 there were six-monthly meetings of the Crossrail and project senior teams to further learn from the delivery of the Crossrail scheme. At this time Andy Mitchell was Programme Director of Crossrail.

Another key way of learning from successful major infrastructure projects was to employ individuals who had worked directly on these schemes. This included a number of key individuals, such as Jim Otta and Peter Shipley. Jim was programme director on a number of deep-tunnel CSO schemes in the USA, as well as the Deep Tunnel Sewage System in Singapore, while Peter was programme director on the London 2012 Olympics. Both had been extremely successful in delivering major infrastructure projects on time and to budget. Our CEO, Andy Mitchell, joined from Crossrail, where he was programme director. He brought with him a wealth of experience in major infrastructure projects, and had masterminded the successful tunnelling phase under the streets of London. Also from Crossrail was Steve Hails, our Health, Safety and Wellbeing (HSW) Director, who'd held the equivalent role there. Andy and Steve were instrumental in developing our thinking and approach to HSW.

Health, Safety and Wellbeing – at the Heart of All We Do

Health and safety is front and centre of everything we do when we are planning, designing and constructing the Thames Tideway Tunnel. It is particularly important on this enormous engineering and construction project, where we are operating on vast scales and in a very challenging environment. This is embodied in our original 'Triple Zero' commitment (aiming for zero incidents, zero harm and zero compromise) and its subsequent evolution into our current absolute commitment to Health, Safety and Wellbeing (HSW).

To ensure we deliver on this commitment, Steve Hails oversees a well-resourced team, this being split across the three project areas in the west, centre and east of the project region, based at the main tunnel drive sites of Carnwath Road Riverside in Fulham, Kirtling Street next to the Battersea Power Station development, and Chambers Wharf in Bermondsey. Each of these teams consists of a lead health and safety manager, an environmental adviser, a health and safety adviser and a Construction, Design and Management (CDM) integrator (supporting the principal designers). Building on this expertise, each of the main works contractors also brings their own HSW teams to work with the Tideway teams.

As outlined above in this chapter, the Tideway HSW team has been fortunate in being able to draw on the previous experiences in this context of the London 2012 Olympics, Heathrow Terminal 2 and Crossrail. In this regard, a key factor going forward regarding HSW on this project has been the realisation that, despite the scale of our programme, its challenges are not unique and we can learn from the experiences of others. This is factored into our daily management of HSW issues.

A central feature of Thames Water and Tideway's commitment to industry-leading health and safety, as mentioned above, is the 'Triple Zero' vision. This is an innovative programme that began life in 2013 when our team held a number of project 'shaping' workshops designed to identify exemplary health and safety best practice in other sectors, such as the nuclear industry, the manufacturing sector and the rail industry. This commitment is aimed at delivering a transformational health and safety performance. As Tideway's previous Head of HSW, Steve Howells, said in 2015:

> If we don't do something fundamentally different to what is currently happening in our industry, we will have fatalities, we will have Reporting of Injuries, Diseases and Dangerous Occurrences Regulations (RIDDORs), we will have high impacts of health incidents, and we will have lots of minor accidents and near misses and that can't be allowed to happen ... Even if we measure ourselves as being a little bit better than what's gone before us that will still happen, so clearly it's not acceptable.[2]

One of the first challenges that the HSW team faced when planning for the beginning of the construction of the Thames Tideway Tunnel was how to avoid the historic spike of incidents that often plague the start of major construction projects, especially those of this size and complexity. Another challenge was how the project team could maintain the focus of all of our Tideway and contractor employees, such that a sustainable active safety (SAS) programme could be built and then maintained for the whole lifetime of the project. In that regard, Howells said in 2015:

> What we are looking to mandate is that every 12 weeks we will stand every site down regardless of the job they do to reflect on where we are but more importantly to do a 24-week look ahead ... We will have a team of safety coaches, safety leaders, and

management looking 24 weeks ahead to say, 'This is where we are going to be in the programme, these are some of the likely hazards we are going to be coming across, so our stand down (see below for explanation) for 24 weeks' time will be focused on that'. We will then have a group who will be developing what that day will look like.[3]

By way of explaining this for the general reader, in a construction context, the term 'stand-down' describes any one of a number of activities where normal work is paused for a certain period of time and the entire site then focuses on a particular safety issue. For the Thames Tideway Tunnel, such stand-downs could last for just an hour or for an entire shift, depending on the stage that the given team is at on the project. We are also clear at Tideway that our wider HSW programme will tie in to national campaigns and health and safety weeks that will be an ongoing feature of our work diary as the Thames Tideway Tunnel project progresses. Our current thinking is that there will be at least four main stand-downs each year across the project, in addition to individual site initiatives. Further, as an additional safety measure, we have decided to enforce the previously mentioned maximum shift length for underground construction workers. Additionally, we have worked hard to understand how best to manage fatigue of the workforce, in particular the effect of shift durations and shift patterns.

We at Tideway have also initiated an immersive induction programme, part of which is held at a purpose-built employee project induction centre known at Tideway by its acronym, EPIC. This programme involves a very early safety engagement process to ensure an understanding of all that Tideway is aspiring to achieve here, and more importantly to get feedback from our HSW teams very early on, helping us to shape and develop the HSW programme based on previous experiences. Part of this induction process involves pre-employment screening to ensure that anyone subsequently employed is authorised to live and work in the UK. There is also a health screening process to ensure that employees are physically fit for the tasks to which they are assigned, especially for safety critical roles and for those at the 'coal face'. As Steve Howells said in 2015:

What we can't do and are not prepared to do is put someone to work who is suffering from a pre-existing condition that we are likely to exacerbate ... if that compromises his or her health and/or (the) safety ... of one of their colleagues, then we can't let that happen.[4]

A further innovation in the pre-induction process is our mandatory health and safety communications assessment. The aim here is to prevent incidents and accidents that would otherwise be caused by workers who have a poor understanding of our health and safety language. For us, the prerequisite for attending EPIC is to take this mandatory test, and then successfully pass it. We have three levels for this test: managers, supervisors and operatives. Successful candidates who pass the assessment and the pre-induction screening then need to complete an online form before being invited through the early safety engagement process to attend the EPIC programme, where they undertake this phase of our immersive multimedia induction programme. The whole process is very experiential and interactive, with up to thirty-six inductees per session. Each day's training focuses on what happens when things go wrong, why they go wrong, and what part we at Tideway may or may not have had to play in the issue in question.

A typical morning session in the EPIC programme consists of a ninety-minute segment during which the attendees gain a clear understanding of what is expected of them when working on the Thames Tideway Tunnel project. The sessions include hazard spotting, reporting and identification, before moving on to cover leadership training. This is to make sure, especially in the case of the operatives, that such skills aren't simply the preserve of managers and supervisors. That is because we believe that true safety leaders are those individuals who make tactical decisions in the face of adversity day in, day out. One of the key elements of this morning session is a presentation by Tideway CEO Andy Mitchell, or one of his directors, for thirty minutes about how to think and act differently in terms of safety to ensure we deliver different outcomes in this regard. The main message is 'if something isn't safe – we don't do it.'

I have undertaken over thirty of the EPIC senior management sessions. At the end of the day, each individual is issued with a project ID card, this matching their biometric hand comparison against their features and face. Further, to ensure even greater levels of security, the ID card only enables the individual to access the sites that they are authorised to visit at any given time on a given day. Everyone leaves with an understanding of what transformational HSW looks like and how we all need to think and behave differently to achieve this objective.

A second half-day in the EPIC programme involves taking workers out on the River Thames to enable them to gain a broader perspective of the project, and also to understand the environmental benefits that we will be delivering (in the same way that external stakeholders are also taken on the river to view the project from this illuminating perspective). Then, with their learnings from the desk-based component of the EPIC programme and their half-day on the river experience, the workers are assigned to the main works contractors to undertake the latter's own on-site inductions, thus further enhancing their training before starting work.

Linked to our focus on healthy workers, we also have an extensive package of wellbeing activities to support our workers when they are not working on our sites. We firmly believe that this final health focus as part of our HSW responsibilities is a key area where we differ from many other such projects of this kind today. For example, working in a congested construction environment, workers may be exposed to health and safety risks not directly related to the construction of the Thames Tideway Tunnel. A case in point is the construction of the Vauxhall Nine Elms regeneration corridor. In this area there are many other major construction projects, such as the redevelopment of Battersea Power Station, the Northern Line Extension and the New American Embassy. In particular, project team members are exposed to health and safety risks associated with vehicle movements on the roads adjacent to our site.

7

PLAN TO REALITY, AND THE FUTURE

The story of the delivery of the Thames Tideway Tunnel, the largest project in the water industry for 150 years and probably for the next 150 years as well, has now passed its halfway point. Construction began in January 2016 and will now progress through to the completion date of early 2024, after which the infrastructure will be handed over to Thames Water as the operator, with Tideway switching at that point to an asset management role. It has been a great success to date, with Mike Gerrard saying: 'Tideway is a world-first in infrastructure investment and shows, yet again, the UK's leadership in this field.'

This is, therefore, a good point in the narrative of the Thames Tideway Tunnel – which reaches its end in terms of my recollections in this chapter – to reflect on the most recent progress made, the key learnings I am keen to pass on to those in the future faced with equally difficult infrastructure projects in the public eye, and the legacy we at Tideway wish to leave to London and its inhabitants through our endeavours.

Bringing the Story Up to Date

The first thing to reference here is that, even today, opposition to the project continues, particularly in relation to a number of individuals. For example, in January 2015 the High Court refused permission to apply for Judicial Reviews with regard to four challenges that had been brought against the Thames Tideway Tunnel DCO process. In three judgments, delivered on 15 and 16 January, Mr Justice Ouseley dismissed all of the claims together with costs. He held that proceedings by the London Borough of Southwark and the Thames Blue Green Plan had been brought out of time. At the same time, claims by the Thames Blue Green Economy Limited were also dismissed as unarguable, given they challenged the designated National Policy Statement that supported the principle of the Thames Tideway Tunnel. Shortly after, in June 2015 the Court of Appeal dismissed the final two Judicial Review proceedings brought against the project. Three of the original four dismissed Judicial Review attempts had been appealed, with one of them earlier being disposed of by a refusal by Lord Justice Sullivan on the papers. As detailed in Chapter 5, however, we remain alert to attempts to frustrate the building of the Thames Tideway Tunnel and continue to be firm in our resolve to maintain momentum in its construction, such that it is completed on time in early 2024.

As with any major infrastructure programmes, the process of designing, consulting and obtaining a DCO, and procuring and engaging stakeholders, has been extensive to say the least. It was, therefore, a truly momentous step when the construction phase finally got going in January 2016 (at the Kirtling Street main tunnel drive site in Battersea, where construction work is now well under way). In the lead up to this significant milestone, we spent over six months setting up our construction operations at the main tunnel drive sites at Carnwath Road Riverside in Fulham, Kirtling Street in Wandsworth and Chambers Wharf in Southwark. Of the twenty-one sites being worked on by Tideway to a greater or lesser extent as part of the construction of the tunnel system, these three are the most important for the overall programme as they are on (or near to) the critical path. They are the manufacturing hubs where the shafts are built and then the TBMs inserted to begin the process of actually boring and constructing the structural lining of the main tunnel. As I write, construction work has commenced on several of our twenty-one sites, including Kirtling Street, Carnwith Road Riverside and Chamber Wharf main tunnel drive sites.

By March 2018, Tideway will be working on nineteen sites across London, a huge ramp-up in activity as we strive to reach our early-2024 completion target. The visibility of the project is set to increase dramatically – particularly at the main tunnel drive sites, where we will be engaged in the construction twenty-four hours a day, seven days a week for around three years, and that is where the Code of Construction Practice (the CoCP detailed in Chapter 5) is particularly important. We are keenly aware that one of the key concerns for those living nearby our construction sites is noise pollution, and to minimise this we are working with electric motors rather than diesels, using acoustic hoods for the machinery, using large warehouse-type structures above the working shafts, and have built up to 5m-high hoardings to surround the sites. For the 700 properties nearest the construction sites, where reasonable noise levels still cannot be guaranteed, we have put in place trigger action plans (TAPs) that allow those living in each of the homes affected to access funds for secondary glazing to their windows and other noise reduction measures.

From a sustainability perspective, one of the key concerns of the stakeholders of all types with whom we have engaged has been ensuring that as much of the transportation of materials for the project as possible (particularly the removal of the excavated material) is undertaken using the river rather than by road. The commitment we made in the project's consent order is that 90 per cent of the excavated material from each of the main drive sites will be taken away by barge (and 56 per cent of all materials moved to/ from the sites), to be reused to provide banking, habitat protection and flood defence further downriver. Through our 'more by river'

Briefing John Manzoni, CEO of the Civil Service, on the project in February 2016.

initiatives, we have worked with the main works contractors to make sure we move 72 per cent of all material by river and Tideway shareholders have invested £45 million to make this commitment possible. It will further reduce the impact of the project on London's already busy roads.

To conclude this short update for the reader, as of September 2017, Tideway as a business has successfully been up and running since 24 August 2015, the construction process is also functioning well, and all of our systems are in place to ensure the smooth operation of the whole business. Further, we continue to engage with our key stakeholders and remain mindful of their concerns, and I truly believe we are making significant progress in building the Thames Tideway Tunnel.

We are far from complacent, however, and continue to innovate in all we do. For example, we have issued challenges to both our contractors and our stakeholders. In particular, we have asked our main works contractors to come up with a delivery programme to take eighteen months off the programme by reducing the consenting and commissioning periods, and starting work on the construction sites even earlier. A number of our stakeholders – i.e. consenting bodies such as the EA and the London local authorities – have been asked to help facilitate this objective.

Although it is unlikely we will deliver the project eighteen months ahead of programme, the aspiration has enabled us to have started construction earlier than intended and to be currently ahead of programme.

Key Learnings from Our Experiences

We have learned many things in bringing the Thames Tideway Tunnel to the point it has reached today. To help share them, these are broken down into the various categories set out below to make them as accessible as possible to the reader, as a key aim of mine has been to pass on lessons learnt from our experiences so that others faced with a similar task are able to benefit from them.

Stakeholder Engagement Lessons

We had a number of learning experiences engaging with stakeholders. For example, as Nick Tennant highlights in Chapter 5, it was very important to present a human face when meeting with those who had views of any kind on the Thames Tideway Tunnel. Even when people opposed us, however aggressively, just being there and being open always had a positive reception, and was much better in helping us present our case than a faceless voice on the end of a telephone. We also found it invaluable to have our experts on hand at all times to help explain any of our activities, both for the stakeholders who clearly appreciated having, for example, the senior engineers explaining to them the workings of the construction process; and also for our team members, as this made the process of promoting the project 'real' when confronted with those affected by it. Nick also says that pitching presentations at the right level was important, and that he found it much more important to be honest and open than slick and over-rehearsed. It is a key competence of the modern engineer to be able to go into the communities affected by a project to sell the need for the scheme, and outline in an open, honest and transparent way the likely disruption and nuisance that will be caused. The project team, myself included, spent thousands of hours in the evenings when residents were available speaking to people face to face. It was a huge commitment.

Another learning experience was the importance of remembering at all times that we (in the private sector) were part of a team, working hand in hand with our public sector partners at the EA, Ofwat and (most importantly) the Government. As Mike Gerrard says, 'Projects like the Thames Tideway Tunnel only happen if numerous private and public sector parties work together.'

In this regard, once the project was specified by the Government as being managed under SIP regulations, Thames Water had a statutory duty to run the IP procurement (eventually won by Tideway) in as open and fair a way as possible. In parallel, Ofwat and the Government also had their own processes to manage in relation to the Project Licence and GSP consultation processes respectively. We therefore found it imperative that the three parties liaised closely and clearly understood their respective roles and processes. In particular, we (Thames Water) learned that it was essential for us to work closely with the Government to develop processes that would result in the selection of a winning bidder (for the IP) in accordance with the relevant procurement law based on an agreed GSP acceptable to the Government. The ultimate manifestation of this teamwork is the Thames Tideway Tunnel Forum for key stakeholders, which for the last five years has met every three months and is now independently chaired. It is a forum that facilitates two-way communication and enables issues to be raised by both parties. I have always had a key role in managing/coordinating these meetings, for many years updating those present on the project's progress.

So the lesson here is clear: go out of your way to work closely and collaboratively with your project partners.

Messaging Lessons

This follows naturally from our stakeholder engagement lessons, in that our failure to initially grasp the importance of how we presented our case to some alienated stakeholders groups quickly became evident.

We initially focused heavily on the fact that we would clean up a very polluted river, to avoid the situation getting worse as London continues to grow and more extreme weather events occur because of climate change. However, as has been stated above a number of times, given the fact the pollution wasn't occuring all the time (only after rainfall events when waste water was released from CSOs), many stakeholders questioned the level of crisis being faced. A common type of response was, 'So you want to spend £4.2 billion to save a few fish?' We quickly learned that a far better approach was instead to focus on capacity. By saying that Bazalgette's system of interceptor sewers was built for a city of 2 million (with a design capacity to cater for a population of 4 million), but was now having to cope with the waste water requirements of a city of over 8 million people and growing, we found that it was much easier for people to grasp the message.

Another mistake, in terms of messaging, was to focus on the tunnel solution as being the only one that could contribute to solving London's future waste water requirements. In so doing we appeared, at least initially, to dismiss any of the 'green' alternatives. This led to us alienating some members of the Green movement, a group you would think would naturally have supported us given our aim is to reduce pollution in the tidal River Thames. Learning from this mistake, in our later engagements with stakeholders we emphasised that the tunnel (while being the key component in meeting future requirements) was actually part of the solution and not the only one. Thus, instead of appearing to dismiss the use of green roofs, swales, permeable pavements and similar, we are now clear that they are also part of the solution. After all, we all want the same outcome: a cleaner and healthier River Thames. The tunnel and SuDS are not mutually exclusive.

We also failed initially to realise how important it was to explain how London's waste water system was interwoven with a wide variety of other (equally complex) underground infrastructure. We solved this by taking key 'opinion formers' down into the Fleet Sewer – this way, they could appreciate the huge infrastructure around, below and through the sewerage system, and therefore understand the scale and complexity of the problem we faced. It was clear to all that it would be virtually impossible to build a separate waste water system: hence the need for a large tunnel solution.

In addition, we focused too much on the technical solution and not on the benefits that the project would bring in terms of legacy. Specifically, this was our 'five pillars of legacy', these being: environment; Health, Safety and Wellbeing; economy; people; and place.

Procurement Lessons

In terms of contractor procurement – essential to the delivery of the whole programme, otherwise the Thames Tideway Tunnel wouldn't be built at all – we benefitted greatly from the incorporation of the best practice developed in other recent exemplar projects in London.

In the first instance, we learned that a well-planned, structured and clearly focused engagement with the main market players in advance of the procurement process helped Thames Water (and later Tideway) to design a packaging strategy and agree on a form of contract that broadly promoted the interests of the project and, as far as possible, appealed to the market place itself. We also determined that there were benefits to be realised during the procurement process, in terms of ensuring there were good lines of communication established with each tenderer. This helped us to understand the dynamics within each tenderer organisation, and to predict accurately the tenderers' reactions to major events during the wider procurement process. Overall, this facilitated the process of generating and maintaining a genuine sense of competitive tension in the process.

Also in terms of procurement, the decision to split the main tunnel into three contracts, and the associated adoption of the WOOR (which ensured that no one company could win more than one contract), was a critical decision that we can now say had far-reaching consequences for the procurement process, and indeed the overall project. To date, this decision appears to have brought significant advantages to the project in terms of maximising competition and also bringing greater commercial resilience to the Thames Tideway Tunnel project.

Land Acquisition Lessons

On a tunnelling project in a major metropolitan area it is absolutely critical to define the sites that are needed for construction and operation of the asset as soon as possible. Once this has been achieved, everything feasible should be done, without delay, to secure the acquisition of the land required. Without the necessary land there is no security that a project can be delivered and, as others have proved, you can end up chasing shadows.

Health, Safety and Wellbeing Lessons

The key learning we had here was don't be afraid or too proud to learn from other projects of a similar scale. Examples include the way we developed our health and safety commitments (through the initial 'Triple Zero' programme), the 'stand-down' procedures to immediately deal with any HSW issue arising, the EaSE early safety engagement process for staff and the EPIC employee induction centre. I think we can be justifiably proud of how we have approached all HSW issues and we would encourage all those faced with similar tasks to learn from the experiences of your industry peers. Our EPIC training is market leading, has won numerous industry awards, and is seen by many as an exemplar of Health, Safety and Wellbeing induction. To date we have had over 9,000 people attend EPIC.

Our Legacy for London and Londoners

The story of the Thames Tideway Tunnel, and Tideway itself, is not simply one regarding the building and operation of the project but also of the wider legacy we will be leaving for future generations. This is neatly summed up by Tideway's chairman, Sir Neville Simms, when he says:

> Creating a cleaner, healthier river is part of our legacy – but it is far from the whole story. We want to create a wider legacy for the capital and, in doing so, help realise the vision to reconnect London with the River Thames. The primary task of the project is to provide the infrastructure that London needs to flourish

economically and socially for decades to come. The intention is to do much more; seizing the opportunity to maximise the broader benefits that the project can offer. A range of benefits will remain that go beyond and outlive tunnel construction, including new standards for health, safety and wellbeing in delivering major projects; a rejuvenated river economy and a new public realm for the people of London to enjoy and connect with the River Thames.

In developing our legacy objectives, we followed four guiding principles: our project must have a positive, lasting effect; a value-for-money solution must be achieved for all objectives; the legacy should raise the standard, building on successes of other nationally significant projects in the UK; and the project should strive for continuous improvement.

Based on these objectives, we can broadly say that the specific overall legacy of our Thames Tideway Tunnel story will be better river water quality and an improved aquatic environment, economic benefits (for example providing for the future of London's essential infrastructure through an enhanced sewerage system that supports growth, and removing the immediate risk of EU imposed infraction fines), jobs (for example with apprenticeship opportunities, where we are creating one for every fifty full-time equivalents), and a river that everybody can enjoy.

Better river water quality and an improved aquatic environment are fundamental to our legacy aims. This is because the clean-up of the River Thames through the project will have a much broader positive impact on life in the capital. The reductions in sewage-related litter will improve the visual appearance of the river, making the river environment look better. This will improve quality of life while reducing costs of collecting the litter for local authorities and others. In addition, the way we build our new structures in the river will provide a legacy of new habitats for aquatic and other wildlife in and around the River Thames. As well as designing our in-river structures with this benefit in mind, we are also taking steps to ensure that land-based wildlife can have greater protection through installing integrated features such as bird and bat boxes. However, we do not want our environmental legacy to be simply based on the physical structures we leave behind. We want to leave behind a better understanding of the river by working with academics and ecologists to provide a broader knowledge. We also want to raise the bar for health and safety performance on future major infrastructure projects as part of our legacy.

Within these specific legacy objectives, however, there are a variety of specific additional legacy aims, these set out in a broad cultural strategy (featuring an underlying theme of 'liberty') which we put in place very early in the narrative of the project.

Helping reconnect London and Londoners with the River Thames

We have this fantastic, iconic river going through the heart of London, which for many years has been heavily polluted by waste water going into it, and we are very keen for Londoners to see the River Thames as an asset that they are proud of, a river they would walk alongside or sit by at their leisure to enjoy.

Putting education at the centre of our activities

We at Tideway have always believed that heritage-related education improves life opportunities, this also being consistent with the 'liberty' theme we have embraced as part of our heritage engagement. Noting that in Chapter 5 I have already set out the importance of engaging with schools as part of our consultation process, we have since then ensured that we use the Tideway story to provide additional resources to educators. This includes the development of Key Stage 2 curriculum teaching resources focused on the social and economic context of local historic buildings, and the development of Key Stage 3 curriculum teaching resources focused on the cultural impact of historic population movements (specifically the long term effect of seventeenth-century Huguenot communities, as Joseph Bazalgette was of Huguenot descent) on the economic, social and urban development of London. Further, the Overarching Archaeological Written Scheme of Investigation (OAWSI) that was part of the planning application process has also provided a wealth of new archaeological and historical information about the River Thames and its environs. This was turned by the Museum of London Archaeology (MOLA) into a Heritage Interpretation Strategy scoping report focused on public engagement, specifically with regard to archaeological interests. It also identified education initiatives that linked the Tideway Science, Technology, Engineering and Mathematics (STEM) education programme with heritage resources located within the River Thames foreshore.

We're also hopeful that, as Bazalgette is so key to our project, the opening of the Thames Tideway Tunnel can encourage new interest in his life and work.

Supporting local heritage projects like Crossness Pumping Station
Here, Tideway is also supporting the Crossness Engines Trust's plans to complete landscaping and the installation of an exhibition exploring the history of the pumping station. The Trust's education objective of opening the historic buildings to visitors complements the Tideway Heritage Interpretation proposals, as it can offer a valuable contribution to the Tideway Heritage Interpretation Strategy (THIS) by specifically examining the history of the Metropolitan Board of Works Main Sewer scheme. Tideway is providing exhibition materials that will assist the Trust in explaining the importance of the site, the role of the Metropolitan Board of Works, the history of urban sanitation and its impact on disease, and the life of Joseph Bazalgette.

As part of our heritage plans, we are also planning to team with partner organisations to make use of additional opportunities to place heritage interpretation displays related to the project within gallery or museum settings across London.

Digital engagement
Given the scale of the information we have had to impart at each stage of the development of the Thames Tideway Tunnel, this has led to us fully embracing digital media for everything from ensuring access to the planning application documents during the DCO application process through to heritage interpretation. Tideway is keen to share our experiences of utilising digital communications, and is already providing digital access via the Tunnelworks (www.tunnelworks.co.uk) platform to an exciting new series of teaching and learning resources that brings the curriculum to life through study of the Thames Tideway Tunnel project.

And so ends the story of the Thames Tideway Tunnel to date. It has been a remarkable journey for me, transformational in many ways and the pinnacle of a well-loved career as an engineer, project manager and director. I know the construction of this complex project is in good hands and will meet all of its programme deadlines and financial objectives, such that by early 2024 Thames Water can look forward to operating a sewerage system fitting for one of the world's leading cities in the twenty-first century. Finally, once again, I would like to say thank you to all of those who have made the Thames Tideway Tunnel a worthy successor to the works of the great Joseph Bazalgette. We are all truly standing on the shoulders of an engineering giant.

The final word goes to Janet Askew, President of the Royal Town Planning Institute, who said of our wonderful project when we won the 2015 Silver Jubilee Cup (the UK's most prestigious planning award) that:

The Thames Tideway Tunnel is an outstanding example of world-class infrastructure being brilliantly delivered with a huge contribution from the planners involved. This worthy winner of our award also impressed with a very positive approach to public engagement, and will result in the vital upgrading of London's sewers that will have an impact that will last for many years to come.

APPENDIX I

Minister of State Ian Pearson MP Letter Instructing the Initiation of the Thames Tideway Tunnel

Nobel House
17 Smith Square
London SW1P 3JR

Telephone 08459 335577
Email mos.environment@defra.gsi.gov.uk
Website www.defra.gov.uk

Department for Environment
Food and Rural Affairs

David Owens
Chief Executive Officer
Thames Water PLC
Clearwater Court
Vastern Road
Reading
RG1 8DB

17 April 2007

From the Minister of State for Climate Change & Environment
Ian Pearson MP

Dear David,

London's sewer overflows

Thank you for your letter of 29 January enclosing additional information on the options for tackling London's sewer overflows. Once again I am most grateful for the efforts Thames Water and others have made to provide this analysis.

I have now considered the additional information you supplied, the Summary Report on Tackling London's Sewer Overflows you sent to me on 29 December 2006, and Regulatory Impact Assessment published on 22 March 2007. On the basis of that information, my view is that an Option 1 type approach – a full-length storage tunnel with additional secondary treatment at Beckton sewage treatment works – is needed. This is both to provide London with a river fit for the 21st century, and for the UK to comply with the requirements of the Urban Waste Water Treatment Directive concerning provision of collecting systems and, in particular, limitation of pollution from storm water overflows.

My view is that early phasing of the Abbey Mills to Beckton tunnel, as well as work on the rest of the scheme, will be needed in order to make progress toward compliance with the Directive (and associated duties under the Water Industry Act and the Urban Waste Water Treatment (England and Wales) Regulations 1994 ('the 1994 Regulations')) as quickly as possible. This is because early phasing of this tunnel would enable 50% of the total volume of collecting system overflow discharges (those to the River Lee from Abbey Mills Pumping Station) to be addressed well before completion of the long tunnel.

I am writing to request that Thames Water makes provision for the design, construction, and maintenance of a scheme for the collecting systems connected to Beckton and Crossness sewage treatment works which:

INVESTOR IN PEOPLE

- involves a full-length storage tunnel with additional secondary treatment at Beckton sewage treatment works;

- meets the requirements of the 1994 Regulations, including for sewerage undertakers to ensure that the design, construction and maintenance of collecting systems is undertaken in accordance with best technical knowledge not entailing excessive costs (BTKNEEC);

- complies with such discharge consent conditions as will be set by the Environment Agency, in exercise of its duty under regulation 6(2) of the 1994 Regulations, to secure the limitation of pollution of the tidal Thames and River Lee due to storm water overflows; and

- limits overflow discharges at Abbey Mills Pumping Station as soon as possible.

I am aware there is further detailed work to be done to design and deliver an appropriate scheme, which from your report may be Option 1c, as quickly as possible. As this will involve major planning, regeneration, funding and financing considerations and applications, I encourage you to continue to work proactively with the relevant parties to identify issues and risks, and how to resolve them. I would be grateful for an Action Plan on these and other key points by Friday 20 April 2007. In the plan I would like to see your outline programme for delivery of a final design, planning and funding applications, and construction of the whole scheme. We plan to share this information with the Commission by the end of April.

From a Government point of view much of the detailed work will fall to Thames Water, the Environment Agency (EA) and the Water Services Regulation Authority (Ofwat) as the environmental and economic regulators. I would be grateful for six monthly progress reports from you on this work. In addition, when important milestones are reached or critical issues arise, such as significant revisions to costs and bill impacts, or concerning the development of plans for Beckton sewage treatment works, I want to be kept informed.

This letter does not amount to enforcement action which would require a precise enforcement order or set of undertakings under sections 18 or 19 of the Water Industry Act 1991. At this stage we do not consider such action to be appropriate, given the further design and feasibility work that needs to be done, or necessary for Thames Water to be able to take matters forward with Ofwat and the Environment Agency.

I am copying this letter to Sir John Harman, Chairman, EA, and Philip Fletcher, Chairman, Ofwat, in order to make them aware of the requirements set out above and of the need for EA and Ofwat to consider appropriate action. I enclose a copy of a letter to Sir John Harman requesting that the EA work with you so as to determine appropriate discharge consent conditions.

Kind regards,

IAN PEARSON

DEFRA letter, to David Owens, CEO Thames Water, dated 17 April 2001, instructing Thames Water to proceed with the single tunnel option.

APPENDIX II

The Shareholders in Bazalgette Tunnel Ltd (Tideway)

Allianz
The Allianz Group is one of the world's leading insurers and asset managers with more than 88 million customers. Allianz holds the leading position for insurers in the Dow Jones Sustainability Index. In 2017, over 140,000 employees in more than seventy countries achieved total revenue of €126 billion and an operating profit of €11 billion. Allianz Capital Partners is the Allianz Group's asset manager for alternative equity investments and manages around €26 billion of alternative assets. The investment focus is on private equity, infrastructure and renewable energy.

Dalmore Capital Limited
Dalmore Capital Limited is an independent UK fund management company with over £5 billion of investors' funds under management and with offices in London and Edinburgh. Dalmore has made around 140 investments in lower volatility infrastructure assets, particularly in the UK. To date, Dalmore's main investment vehicle is PPP (Public–Private Partnership) Equity PIP LP. For the purpose of investment in the Thames Tideway Tunnel, Dalmore has established a single purpose fund that has secured £440 million of commitments, primarily from leading UK pension funds as well as a small number of European investors. Alistair Ray, Partner at Dalmore (and co-founder/Chief Information Officer), said regarding the Tideway investment:

> To date it has been difficult for UK pension funds to invest into major UK infrastructure projects such as Thames Tideway Tunnel where overseas funds have dominated. We hope that following this structure we can similarly bring UK investors into other high profile transactions with similar low risk inflation linked returns. Dalmore believes this is good news for both the consumer and UK Plc.

DIF
DIF is a leading independent infrastructure fund manager, with €5.6 billion of assets under management across multiple closed-end infrastructure funds. DIF invests in greenfield and brownfield

197

infrastructure assets that have long duration, stable and predict-able cashflows, located primarily in Europe, North America and Australasia. This includes public-private partnerships, concessions, regulated assets, renewable energy projects and contracted assets. Since inception in 2005, DIF has invested in over 190 infrastructure assets with a total value of over €30 billion.

International Public Partnerships Limited (INPP)
INPP is a listed infrastructure investment company that invests in high-quality, predictable, long-duration public infrastructure pro-jects and businesses in the UK, Australia, Europe and Canada. Listed in 2006, INPP's portfolio comprises over 130 investments across a variety of sectors including social infrastructure, environmental infrastructure, transport, energy transmission and distribution, and digital infrastructure.

Swiss Life Asset Managers
Swiss Life Asset Managers has more than 160 years of experience in managing the assets of the Swiss Life Group. This insurance background has exerted a key influence on the investment phi-losophy of Swiss Life Asset Managers, which is governed by such principles as value preservation, the generation of consistent and sustainable performance and a responsible approach to risks. As of 31 December 2017, Swiss Life Asset Managers managed a total volume of over €69 billion assets for the Swiss Life Group.

Amber Infrastructure Group
Amber Infrastructure Group is a leading international infrastructure specialist, providing asset management and investment advisory services in respect of over £8 billion of assets in the UK, Europe, Australia and North America. Amber's core business focuses on sourcing, developing, advising on, investing in and managing infra-structure assets including regulated utilities, social infrastructure and sustainable energy. Amber provides investment advisory ser-vices to International Public Partnerships Limited as well as private investment funds, specialising in urban regeneration and digital infrastructure. Amber is based in London, with offices in Munich, Sydney and San Francisco, and employs over 120 people, making it one of the largest international infrastructure specialists.

APPENDIX III

River Thames Combined Sewer Overflows

	Thames Tideway Tunnel CSOs	
CSO ID	CSO Name	Method of Control
CS01X	Acton Storm Relief	Intercepted (at site ACTST)
CS02X	Stamford Brook Storm Relief	Indirectly controlled
CS03X	North West Storm Relief	Indirectly controlled
CS04X	Hammersmith Pumping Station	Intercepted (at site HAMPS)
CS05X	West Putney Storm Relief	Intercepted (at site BAREL)
CS06X	Putney Bridge	Intercepted (at site PUTEF)
CS07A	Frogmore Storm Relief (Bell Lane)	Intercepted (at site DRMST)
CS07B	Frogmore Storm Relief (Buckhold Road)	Intercepted (separated site from CS07A at KNGGP)
CS08A	Jews Row (Wandle Valley Storm Relief)	Indirectly controlled
CS08B	Jews Row (Falconbrook Storm Relief)	Indirectly controlled
CS09X	Falconbrook Pumping Station	Intercepted (at site FALPS)
CS10X	Lots Road Pumping Station	Intercepted (at site CREWD)

CS11X	Church Street	Indirectly controlled
CS12X	Queen Street	Indirectly controlled
CS13A	Smith Street (Main Line)	Indirectly controlled
CS13B	Smith Street (Storm Relief)	Indirectly controlled
CS14X	Ranelagh	Intercepted (at site CHEEF)
CS15X	Western Pumping Station	Indirectly controlled
CS16X	Heathwall Pumping Station	Intercepted (at site HEAPS)
CS17X	South West Storm Relief	Intercepted (same site as CS16X)
CS18X	Kings Scholars Pond	Indirectly controlled
CS19X	Clapham Storm Relief	Intercepted (at site ALBEF)
CS20X	Brixton Storm Relief	Intercepted (same site as CS19X)
CS21X	Grosvenor Road	Indirectly controlled
CS22X	Regent Street	Intercepted (at site VCTEF)
CS23X	Northumberland Street	Indirectly controlled
CS24X	Savoy Street	Indirectly controlled
CS25X	Norfolk Street	Indirectly controlled
CS26X	Essex Street	Indirectly controlled
CS27X	Fleet Main	Intercepted (at site BLABF)

CS28X	Shad Thames Pumping Station	Sewerage modification (at site SHTPS)
CS29X	North East Storm Relief	Intercepted (at site KEMPF)
CS30X	Holloway Storm Relief	Sewerage modification (at site BEKST)
CS31X	Earl Pumping Station	Intercepted (at site EARPS)
CS32A	Deptford Storm Relief	Intercepted (at site DEPCS)
CS32B	Deptford Creek	Indirectly controlled
CS33X	Greenwich Pumping Station	Intercepted (at site GREPS)
CS34X	Charlton Storm Relief	Indirectly controlled
Lee Tunnel CSO		
CS35X	Abbey Mills Pumping Station	Intercepted (at site ABMPS)
Remaining CSOs		
CS36X	Wick Lane	Controlled by separate project
CS81X	Tideway CSO	New CSO at B
CS37X	Northern Low Level 1-Brook Green	Influenced
CS39X	Horseferry	Influenced
CS40X	Wood Street	Influenced
CS41X	Goswell Street	Influenced
CS42X	Pauls Pier	Influenced
CS43X	Battle Bridge	Influenced
CS44X	Beer Lane	Influenced
CS45X	Iron Gate	Influenced
CS46X	Nightingale Lane	Influenced

CS47X	Union Wharf	Influenced
CS48X	Wapping Dock	Influenced
CS49X	Cole Stairs	Influenced
CS50X	Bell Wharf	Influenced
CS51X	Ratcliffe	Influenced
CS52X	Blackwall Sewer	Influenced
CS53X	Henley Road	Influenced
CS54X	Store Road	Not in service
CS55X	London Bridge	Influenced
CS56X	Isle of Dogs Pumping Station	Influenced
CS57X	Canning Town Pumping Station	Not in service

APPENDIX IV

Principal Geological Features Along the Length of the Thames Tideway Tunnel Project

The dominant structural geological features that impact the Thames Tideway Tunnel project are:

The Hammersmith Reach Fault Zone, a series of north-north-west to south-south-east trending faults beneath and adjacent to the eastern side of Hammersmith Bridge. A 5m displacement features to the east.

The Putney Bridge Fault, a series of south-east to north-west trending faults on the syncline with the axis to the west of Putney Bridge, with vertical displacement on top of Lambeth Group strata on the eastern hanging wall of approximately 2m.

The Chelsea Embankment (Albert Bridge) Fault Zone, a series of north to south and south-south-west to north-north-east trending faults between Battersea and Chelsea bridges, which intersect the main tunnel alignment at nearly perpendicular. Up to 5m vertical displacement of strata is a feature of this zone, resulting in uplift of the top of Lambeth Group deposits on the eastern side of Albert Bridge.

The Lambeth Anticline, a north-north-west to south-south-east trending faulted anticline between Vauxhall and Lambeth Bridges that intersects the main tunnel alignment at an oblique angle with a difference in strata level of approximately 5m.

The London Bridge Fault Graben (a depressed block of land which is bordered by parallel faults), a south-east to north-west trending graben-type feature arranged between Cannon Street and Tower Bridge, with known vertical displacements in excess of 10m.

The Greenwich Fault Zone, a south-west to north-east trending feature, which was investigated in detail in 2008 as part of the Lee Tunnel project ground investigation. Up to 20m down-throw is a feature here to the north-west in a series of stepped faults. The fault runs generally parallel to the main syncline, south-west to north-east from Greenwich to Beckton, crossing the River Thames downstream of the Thames Barrier. It is in close proximity to Greenwich Pumping Station.

Other structural features include **the North Greenwich Syncline** (now more generally known as the Plaistow Graben), **the Millwall Anticline** and **Beckton Anticline**, all of which trend north-east to south-west, contrary to the main basin axis.

Meanwhile, during the construction process of the main tunnel, there are also risks from scour hollows, located on previous drainage channels and formed by the River Thames at the confluence of the existing tributaries (notably the Rivers Fleet, Lee and Wandle). They usually feature a variety of granular deposits and/or disturbed natural materials and are localised and steep-sided. The scour hollow in the vicinity of the Blackwall Tunnel is the only one known to penetrate into the natural chalk. Elsewhere, the hollows only affect the tertiary deposits and, more particularly, the London Clay. Such basal depths are normally between 5m to 20m below ground level, with the exception of 33m at Battersea Power Station and Hungerford Bridge.

Of the known scour hollows, only the hollow at Hungerford Bridge is close to the main tunnel. This feature attains a base level of 72m ATD (-25m AOD) in London Clay near the south bank, equivalent to only 10m above the main tunnel crown. Such features may, however, have greater implications for the shallower connection tunnels in other locations (see Chapter 6 for detail of the various tunnel types being used in the wider Thames Tideway Tunnel system).

Meanwhile, the presence of flints within the chalk may cause severe wear to the tunnel-boring machines in areas where this rock is a significant feature, which will require frequent and hazardous interventions to inspect and maintain the TBM cutterhead. Therefore, an important part of the project's ground investigations prior to submitting the development consent application was to investigate the structure and permeability of the chalk and the characteristics of flint-related features. In terms of their locations, a number of flint bands are present within the chalk. For example, within the Seaford chalk, the two well-defined flint bands used as marker horizons (not necessarily the thickest seams) are the Bedwells Columnar and Seven Sisters. The Bedwells typically comprise a discontinuous layer of very large, irregular flints, up to approximately 500mm in height by 300mm in diameter. Previous projects have determined that they have a compressive strength of up to 900MPa. Meanwhile, the Seven Sisters is a continuous band, with flints between 100mm and 150mm thick. It is therefore important to select the appropriate TBMs for this task; for example, a slurry TBM is preferred for the section of the route in chalk. By way of example, a slurry TBM was used successfully on the Channel Tunnel Rail Link River Thames crossing next to the QE2 Bridge. Most recently, the same type of TBM was procured by the Lee Tunnel project contractor. The advantages of this type of TBM includes the ability to deal with water-bearing fissures in chalk and to convey flint pieces in a fluid slurry, rather than a potentially damaging abrasive paste from an Earth Pressure Balance (EPB) TBM. Selecting a Slurry TBM for Chalk also reduces the need for hazardous interventions. On the downside, however, Slurry TBMs are not appropriate for London Clay of the type expected along other sections of the main tunnel alignment, hence as set out in Chapter 6 the need to use multiple different types of TBM for the project.

Geotechnical investigation jack-up barge undertaking borehole work in the River Thames next to Tower Bridge.

Inspecting one of the borehole cores from the geotechnical investigation at the store at Abbey Mills Pumping Station, March 2010.

APPENDIX V

Hydrogeology and the Thames Tideway Tunnel

The major aquifer of the London Basin lies in the chalk beds. It is wholly unconfined to the east, but confined to the west below the tertiary strata (and the London Clay Formation in particular). The chalk aquifer is generally in hydraulic continuity with the overlying Thanet Sand Formation and sometimes the base of the Lambeth Group, particularly the gravel part of the Lower Mottled Beds and the Upnor Formation. The EA refers to this combined aquifer as the Chalk-Basal Sands aquifer. Local aquicludes (impermeable bodies of sediment stratum, or rock, that act as a barrier to the flow of groundwater) can exist in the overlying Lambeth Group, in particular the Woolwich Formation Laminated Beds, this leading to perched groundwater tables.

Historical records of other engineering schemes state that these 'perched' features may retain hydrostatic pressures of up to 40m, which could result in high inflows at tunnel level and particularly into shafts during construction. The Harwich Formation (Blackheath Member) is also known to contain high groundwater levels in places, which may cause problems during tunnel construction. A minor regional aquifer also lies within the floodplain and river terrace deposits. Due to its connection to the River Thames, it is generally tidal.

APPENDIX VI

Consultation Process Schedule

Phase	Start	End	Number of Weeks / Days	Mailing
Phase One consultation	13 September 2010	14 January 2011	18 weeks	over 162,000 letters
Interim engagement	11 March 2011	16 August 2011 Plus Kirtling Street site, which was undertaken between 10 and 11 October 2011.	22 weeks	Mail drops to areas within 250 metres
Phase Two consultation	4 November 2011	10 February 2012	14 weeks	129,000 letters
Post-Phase Two consultation (Target consultation)				
Barn Elms, Putney Embankment Foreshore, Albert Embankment Foreshore and Victoria Embankment Foreshore Abstract water or operate ground source heat pumps	6 June 2012	4 July 2012	29 days	Over 15,000 letters
Changes to site boundaries at Albert Embankment Foreshore and Falconbrook Pumping Station	2 August 2012	5 October 2012	65 days	17 letters
Changes to site boundaries at King George's Park and Cremorne Wharf Depot	26 October 2012	26 November 2012	32 days	Over 40 letters

Sensitive Equipment	29 October 2012	26 November 2012	28 days	Over 450 letters
Section 48 Publicity	16 July 2012	5 October 2012	12 weeks	Statutory Stakeholders only
Targeted Engagement Victoria Embankment Foreshore and Blackfriars Bridge Foreshore	15 July 2013	12 August 2013	28 days	Over 50 letters
Targeted Engagement Victoria Embankment Foreshore and Blackfriars Bridge Foreshore	4 October 2013	12 November 2013	40 days inc w/e	Over 740

APPENDIX VII

Particular Thanks

Note: Thames Tideway Tunnel project positions are detailed as at the time of writing or the positions held when the individual left the project.

Amar Qureshi, Commercial Director, Thames Tideway Tunnel, TWUL
Andy Mitchell, CBE, CEO, Bazalgette Tunnel Ltd
Anne Richards, Property Manager, CH2M
Charlotte Morgan, Partner, Linklaters
Chris Boston, Property Manager, TWUL
Chris Stratford, Planning Manager, CH2M
Clare Donnelly, Lead Architect, Fereday Pollard Architects
Clare Gibbons, Planning Manager, TWUL
Dave Wardle, London Area Manager, EA
David Crawford, Chief Engineer of System Design and Operations, CH2M
Derek Arnold, West Engineering Manager , CH2M
Gareth Thomas, Central Engineering Manager, CH2M
Ian Fletcher, Planning Manager, Bazalgette Tunnel Ltd
Ian Pearson, Minister of State for Climate Change and Environment, DEFRA (2006–07)
Ingrid Lagerberg, Systems Engineering Lead, CH2M
James Good, Partner BLP
Jeff Meerdink, Head of Consents Authority, CH2M
Jim Fitzpatrick, MP for Poplar and Limehouse
Jim Otta, Programme Director, CH2M (2008–15)
John Bourne, Deputy Director, Water Availability and Quality, DEFRA
John Greenwood, Design Manager, TWUL
John Ramage, Project Director, CH2M
John Rhodes OBE, Director, Quod
Justine Greening, MP for Putney
Keith Mason, Senior Director Finance and Networks, Ofwat
Kevin Reid, Principal Programme Manager, GLA
Lord de Mauley, Parliamentary Under-Secretary, DEFRA (2012–15)
Lord Deighton, Commercial Secretary, HM Treasury (2013–15)

Malcolm Orford, Deputy Delivery Manager (Central), CH2M
Martin Baggs, CEO TWUL
Michael Humphries QC, Francis Taylor Building
Mike Gerrard, Managing Director, Thames Tideway Tunnel, TWUL
Nick Butler, Project Sponsor, Bazalgette Tunnel Ltd
Nick Fincham, Strategy and Regulation Director TWUL
Nick Sumption, Industry Innovation Lead, Bazalgette Tunnel Ltd
Nick Tennant, Head of Corporate & Public Affairs, Bazalgette
    Tunnel Ltd
Nicky Sumner, Partner, Sharpe Pritchard
Patricia Stevenson, Site Selection Manager, AECOM
Peter Shipley, Programme Director, CH2M
Professor Chris Binnie, former Chair, TTSS
Richard Aylard, CVO, External Affairs and Sustainability Director,
    TWUL
Richard Benyon MP, Parliamentary Under-Secretary, DEFRA
    (2010–13)
Richard Threlfall, Partner, KPMG
Rick Fornelli, Systemwide Delivery Manager, CH2M
Roger Bailey, Asset Management Director, Bazalgette Tunnel Ltd
Sian Thomas, Head of Asset Management, Bazalgette Tunnel Ltd
Simon Hughes, MP for Bermondsey and Old Southwark (1983–2015)
Sir Neville Simms, Chairman, Bazalgette Tunnel Ltd
Sonia Phippard, Director Water and Flood Risk management, DEFRA
Stephen Dance, Head of Infrastructure Delivery, Infrastructure UK
Stephen Paine, Managing Director, UBS
Steve Walker, Head of Thames Tideway Tunnel Project, Ofwat
Stuart Siddall, CFO TWUL
Sue Hitchcock, Commercial Manager, Bazalgette Tunnel Ltd
Suzanne Burgoyne, Environment Manager, Bazalgette Tunnel Ltd
William Lambe, Finance Director, Thames Tideway Tunnel, TWUL

APPENDIX VIII

The Lost Rivers of London

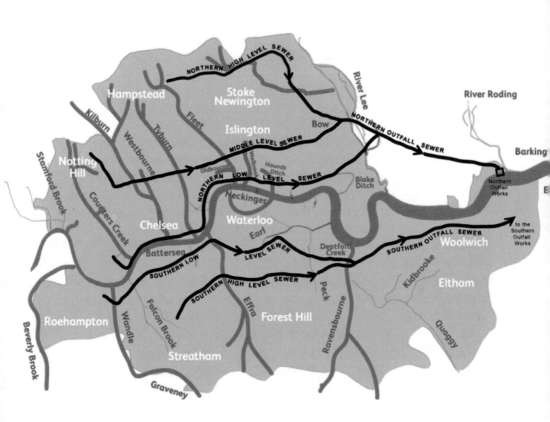

LIST OF REFERENCES

Books

Ackroyd, P. 2012. London Under. London: Chatto and Windus.
Darlington, I. 1970. *The London Commissioners of Sewers and Their Records*. Chichester: Phillimore & Company Ltd.
Elliott, S. 2016. *Sea Eagles of Empire: The Classis Britannica and the Battles for Britain*. Stroud: The History Press.
Forte, A., Oram, R., and Pedersen, F. 2005. *Viking Empires*. Cambridge: Cambridge University Press.
Halliday, S. 1999. *The Great Stink of London: Sir Joseph Bazalgette and the Cleansing of the Victorian Metropolis*. Stroud: Sutton Publishing.
Overy, R. 2014. *The Bombing War*. London: Penguin.
Ross, C. 1976. *The Wars of the Roses*. London: Thames and Hudson.
Schofield, J. 2009. *Cromwell to Cromwell*. The History Press: Stroud.
Thomas, C. 2003. *London's Archaeological Secrets Revealed*. New Haven: Yale University Press.
Tomlin, S.O. 2016. *Roman London's First Voices*. London: MOLA.

Articles and Reports

Cohen, N. 2010. 'Liquid History: Excavating London's Great River, the Thames.' *Current Archaeology*, Issue 144, V.20, 10.
Institution of Civil Engineers. 1991. 'Sir Joseph Bazalgette. Civil Engineering in the Victorian City.' London: Institution of Civil Engineers Archives Panel.
Pump, C. 2011. *Air Entrapment Relationships with Water Discharge of Vortex Drop Structures*. MSc thesis. Iowa City: University of Iowa – Iowa Research Online. Available from: ir.uiowa.edu/cgi/viewcontent.cgi?article=2447&context=etd [Accessed 18 December 2016].
Thames Tideway Strategic Study. 'Steering Group Report.' Available from: thameswater.co.uk/cps/rde/xbcr/corp/ttss-background-steering-group-report.pdf [Accessed 28 December 2016].
Thames Water. 2016a. 'Thames Tideway Tunnel IP Procurement Report.' Unpublished.

Thames Water. 2016b. 'Thames Tideway Tunnel Main Works Procurement Report.' Unpublished.

Thames Water. 2016c. 'Thames Tideway Tunnel Overarching Procurement Report.' Unpublished.

Thames Water. 2013. 'Application for Development Consent Ref 7.18– Engineering Design Statement.' Available from: tideway.london

Thames Water. 2013. 'Application for Development Consent Ref 7.05 – Final Report on Site Selection.' Available from: tideway.london

Thames Water. 2013. 'Thames Tideway Tunnel Air Management Plan. Application for Development Consent Ref 7.14'

Online

Essex Chambers. January 2015. 'Final Thames Tideway Tunnel Judicial Reviews Dismissed.' Available from: www.39essex.com/judicial-reviews-thames-tideway-tunnel-dismissed [Accessed 25 November 2016].

Brockett, J. 2016. 'Interview: Phil Stride, Strategic Projects Director, Tideway. Water and Waste Water Treatment.' Available from: wwtonline.co.uk/features/interview-phil-stride-strategic-projects-director-tideway [Accessed 18 October 2016].

Blue Green UK. Available from: bluegreenuk.com [Accessed 14 December 2016].

Blue Green UK, 'Thames Blue Green Economy.' Available from: bluegreenuk.com/freewater/tbge.html [Accessed 8 October 2016].

CH2M Hill. 'Transforming the Ecology of London's River.' Available from: www.ch2m.com [Accessed 19 November 2016].

Courtney, A. 2012. 'Sewer Protestors Gather Momentum in Push for Consultation Extension.' in *getwestlondon*. Available from: www.getwestlondon.co.uk/news/local-news/sewer-protesters-gather-momentum-push-5980922 [Accessed 27 October 2016].

Department of Agriculture, Environment and Rural Affairs, Northern Ireland Assembly. 'Urban Waste Water.' Available from: www.daera-ni.gov.uk/articles/urban-waste-water [Accessed 10 December 2016].

Department for Environment, Food and Rural Affairs. 2007. 'Background towards a solution to combined sewer overflows affecting water quality in the Thames Tideway and River Lee: Thames Tideway Strategic Study.' Available from: webarchive.nationalarchives.gov.uk/20130123162956/http:/www.defra.gov.uk/environment/quality/water/waterquality/sewage/documents/overflows-background.pdf [Accessed 10 November 2016].

Department for Environment, Food and Rural Affairs. 2014. 'The Thames Tideway Tunnel Preparatory Works Notice.' Available from: www.gov.uk/government/uploads/system/uploads/attachment_data/file/317560/TTTP-prep-work-reason-notice-ldmsig.pdf [Accessed 11 September 2016].

Department for Environment, Food and Rural Affairs. 2014. 'The Water Industry (Specified Infrastructure Projects) Regulations 2013 – Thames Tideway Tunnel Project Specification Notice.' Available from: www.gov.uk/government/uploads/system/uploads/attachment_data/file/317556/TTTP-notice-ldmsig.pdf [Accessed 9 October 2016].

Environment Agency. 2015. 'Fish Survey Reveals Huge Variety in River Thames.' Available from: www.gov.uk/government/news/fish-survey-reveals-huge-variety-in-river-thames [Accessed 10 October 2016].

Excell, J. 2014. 'Sewage Solution.' *The Engineer.* (Online). Available from: www.theengineer.co.uk/issues/february-digital-issue/sewage-solution/ [Accessed 14 September 2016].

Infrastructure Intelligence. 2015. 'Thames picks preferred bidder to finance, build and own Tideway.' Available from: www.infrastructure-intelligence.com/article/jul-2015/thames-picks-preferred-bidder-finance-build-and-own-tideway [Accessed 9 November 2016].

Lanktree, G. 2015. 'Thames Sewage: UK Faces up to Prospect of Multi-million Pound EU Fines'. *International Business Times.* Available from: www.ibtimes.co.uk/thames-sewage-uk-faces-prospect-multimillion-pound-eu-fines-1532945 [Accessed 8 August 2016].

Mulholland, H. 2009. 'UK Faces European Court for Allowing Raw Sewage to Enter Thames'. *The Guardian.* Available from: www.theguardian.com/politics/2009/oct/09/river-thames-pollution-european-union [Accessed 20 December 2016].

New Civil Engineer. 'CH2M Hill to Manage Thames Tideway Tunnel.' Available from: www.newcivilengineer.com/ch2m-hill-to-manage-thames-tideway/945276.article [Accessed 27 October 2016].

Ofwat. 2015. 'Thames Tideway Tunnel Infrastructure Provider Licence.' Available from: www.ofwat.gov.uk/regulated-companies/improving-regulation/thames-tideway [Accessed 20 September 2016].

Petitionbuzz: 2013. 'Stop Sewer Shaft in Carnwath Road Fulham'. Available from: www.petitionbuzz.com/petitions/stopfulhamsewershaft [Accessed 15 November 2016].

Save Our Riverside. 2014. 'SYR Condemns Dreadful Tunnel Decision'. Available from: www.saveyourriverside.org [Accessed 17 October 2016].

SuDs Wales. 'SuDs Techniques – Permeable Conveyance Systems.' Available from: www.sudswales.com/types/permeable-conveyance-systems/swales [Accessed 16 December 2016].

Thames Water. 'Tackling London's Sewer Overflows.' Available from: infrastructure.planninginspectorate.gov.uk/wp-content/ipc/uploads/projects/WW010001/WW010001-001266-8.2.3_Objectives_and_Compliance_Working_Group_Report_Volume_2.pdf [Accessed 19 December 2016].

United Nations. 2014. 'World's population increasingly urban with more than half living in urban areas.' Available from: www.un.org/en/development/desa/news/population/world-urbanization-prospects-2014.html [Accessed 10 August 2016].

UK Construction Online. 2015. 'Thames Tunnel Tideway contracts Announced'. Available from: www.ukconstructionmedia.co.uk/news/thames-tideway-tunnel-contracts-announced [Accessed 1 December 2016].

Warburton, N. 2015. 'A Vision of Safety at Thames Tideway Tunnel'. *SHP Online.* Available from: www.shponline.co.uk/thames-tideway-tunnel [Accessed 20 November 2016].

NOTES

Chapter 1

1 Thomas, C. 2003. *London's Archaeological Secrets Revealed.* p.18
2 Elliott, S. 2016. *Sea Eagles of Empire: The Classis Britannica and the Battles for Britain.* p.109.
3 Forte, A., Oram, R., and Pedersen, F. 2005. *Viking Empires.* p.76.
4 Ross, C. 1976. *The Wars of the Roses.* p.64.
5 Schofield, J. 2009. *Cromwell to Cromwell.* p.175.
6 Overy, R. 2014. *The Bombing War.* p.126.
7 Cohen, N. 2010. 'Liquid History: Excavating London's Great River, the Thames.' *Current Archaeology*, Issue 144. p.10.
8 Thomas, *Ibid.* p.18.
9 Tomlin, S.O. 2016. *Roman London's First Voices.* p.15.

Chapter 2

1 Ackroyd, P. 2012. *London Under.* p.19.
2 Darlington, I. 1970. *The London Commissioners of Sewers and Their Records.* p.3.
3 *Ibid.*
4 Halliday, S. 1999. *The Great Stink of London: Sir Joseph Bazalgette and the Cleansing of the Victorian Metropolis.* p.24.
5 *Ibid.* p.25.
6 *Ibid.* Preface.
7 *Ibid.*
8 *Ibid.* p.72.

Chapter 3

1 Mulholland, H. 2009. 'UK Faces European Court for Allowing Raw Sewage to Enter Thames'. *The Guardian.*
2 *Ibid.*

3 *Ibid.*
4 Lanktree, G. 2015. 'Thames Sewage: UK Faces up to Prospect of Multi-million Pound EU Fines'. *International Business Times.*
5 As detailed by DEFRA in Department for Environment, Food and Rural Affairs. 2007. 'Background towards a solution to combined sewer overflows affecting water quality in the Thames Tideway and River Lee: Thames Tideway Strategic Study.'

Chapter 4

1 Thames Water. 2016a. 'Thames Tideway Tunnel IP Procurement Report.'
2 *Ibid.*
3 *Ibid.*
4 *Ibid.*
5 Department for Environment, Food and Rural Affairs. 2014. 'The Thames Tideway Tunnel Preparatory Works Notice.'
6 Department for Environment, Food and Rural Affairs. 2014. 'The Water Industry (Specified Infrastructure Projects) Regulations 2013 – Thames Tideway Tunnel Project Specification Notice.'

Chapter 5

1 See Appendix A.
2 Courtney, A. 2012. 'Sewer Protestors Gather Momentum in Push for Consultation Extension.' *getwestlondon.*
3 Lanktree, *Ibid.*

Chapter 6

1 Pump, C. 2011. *Air Entrapment Relationships with Water Discharge of Vortex Drop Structures.* MSc thesis. Iowa City: University of Iowa.
2 Warburton, N. 2015. 'A Vision of Safety at Thames Tideway Tunnel'. *SHP Online.*
3 *Ibid.*
4 *Ibid.*

INDEX

Note: *italicised* numbers denote illustrations.

You may also enjoy …

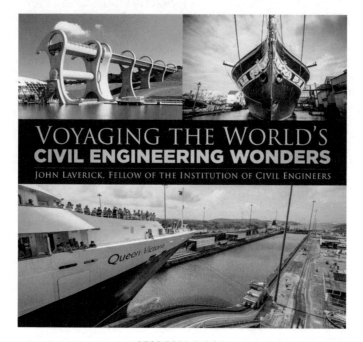

9780 7509 8436 2

John Laverick takes you on a journey
that crosses three continents and three
centuries, exploring extraordinary
achievements including the artificial
waterways of the Panama and Suez
canals, the world's only rotating ship
lift at Falkirk, and a man-made island in
the Baltic linking the crossings between
two countries.